AN ILLUSTRATED GUIDE TO LINEAR PROGRAMMING

by Saul I. Gass

College of Business and Management
University of Maryland

Illustrated by W. F. McWilliam

DOVER PUBLICATIONS, INC., New York

Published in Canada by General Publishing Company, Ltd., 30 Lesmill Road, Don Mills, Toronto, Ontario.

Published in the United Kingdom by Constable and Company, Ltd., 10 Orange Street, London WC2H 7EG.

This Dover edition, first published in 1990, is an unabridged, slightly corrected republication of the work first published by the McGraw-Hill Book Company, New York, 1970.

Manufactured in the United States of America
Dover Publications, Inc., 31 East 2nd Street, Mineola, N.Y. 11501

Library of Congress Cataloging-in-Publication Data

Gass, Saul I.
An illustrated guide to linear programming / by Saul I. Gass ; illustrated by W. F. McWilliam.
p. cm.
Reprint. Originally published: New York : McGraw-Hill, 1970.
Includes bibliographical references.
ISBN 0-486-26258-8
1. Linear programming. I. Title.
T57.74.G28 1990 89-25752
519.7′2—dc20 CIP

Do not try to satisfy your vanity by teaching a great many things. Awake their curiosity. It is enough to open their minds, do not overload them. Put there just a spark. If there is some good inflammable stuff, it will catch fire.

ANATOLE FRANCE

To Ron and Joyce,
who have the spark.

PREFACE

The beginnings of all things are small.
CICERO

INTEREST IN SCIENTIFIC MANAGEMENT and the related arsenal of technical weaponry encompasses a rather wide and ever-increasing audience. Since the field of linear programming has been a key element in the burgeoning power of scientific management, I felt it appropriate to prepare an elementary book which would serve this audience in a multipurpose sense. For some, it is hoped that the *Guide* will act as a catalyst and send them off to deeper presentations and related areas. For others, this *Guide* should be adequate for their needs. For all readers, it is hoped that an hour or two with the *Guide*—with some or little thought about the limited mathematical discussions—will enable them to garner the essentials of linear programming. Thus, *An Illustrated Guide to Linear Programming* is designed to present the rudiments of a complex topic in an elementary, but informative and entertaining fashion.

As I have been actively involved in the field of linear programming since 1952, it is rather difficult to discriminate and delineate the numerous influences which have shaped the presentation of the *Guide*. Its origins can be traced to a nontechnical presentation I authored in 1954 while a member of the Directorate of Management Analysis, U.S. Air Force. But through the years, I have gained in knowledge and concepts via readings, lectures, meetings, bull-sessions, and, of course, many stimulating discussions with friends and associates. Their subliminal contributions are gratefully acknowledged.

I want to express my deep appreciation to my wife, Trudy, for her continuing encouragement and patience, to Bill McWilliam for capturing the essence of the subject matter in his excellent illustrations, and to Joanne Wagner for her accomplishing the most difficult task of reading my pen scratches and translating them to typed copy.

Saul I. Gass

CONTENTS

AN ILLUSTRATED GUIDE TO LINEAR PROGRAMMING

INTRODUCTION

{On how to get dressed, about straight lines, and the beginning of a trip through the land of Linear Programsville.}

THE SEARCH for the best solution—the maximum, the minimum, or in general, the optimum solution—to a wide range of problems has entertained and intrigued man throughout the ages. Euclid described ways to find the greatest and least straight lines that can be drawn from a point to the circumference of a circle, and how to determine the parallelogram of maximum area with given perimeter. The

great mathematicians of the seventeenth and eighteenth centuries developed new optimization procedures that solve complex geometric, dynamical, and physical problems, such as finding the minimum curves of revolution or the curve of quickest descent.

Recently, a new class of optimization problems has originated out of the convoluted organizational structures that permeate modern society. Our natural inclination to attack and solve such problems is manifested by such phrases as "cost-effective," "mostest for the leastest," and "more bang for the buck." Here we are concerned with such matters as the most efficient manner in which to run an economy or a factory, the optimum deployment of aircraft that maximizes a country's chances for winning a war, or with such mundane tasks as mixing cattle feed to meet diet specifications at minimum cost. Research on how to formulate and solve such problems has led to the development of new and important optimization techniques. Among these we find the subject of this book—*linear programming.*

To define linear programming in nontechnical terms we can take two approaches—describe the literal meaning of the phrase linear programming or simply describe typical problems which can be formulated as linear programs. As it is quite instructional to interweave both descriptive paths, we shall do so.

In a most general sense, programming problems—linear or otherwise—are concerned with the efficient use or allocation of limited resources to meet desired objectives. Such allocation problems are central to the field of economics. However, they not only are found within industrial and corporate entities, but also arise in many guises during an individual's encounter with his day's activities.

ON GETTING DRESSED

Diddle, diddle, dumpling, my son John,
He went to bed with his stockings on;
One shoe off, one shoe on;
Diddle, diddle, dumpling, my son John.
ANONYMOUS

The first thing each morning you and I face the programming problem of getting dressed. We must select a program of action which enables us to become dressed in a manner which meets the constraints or accepted fashion rules of society—socks are not worn over shoes, but socks are worn. Our basic resource is time, and the selected program must be best in terms of how each individual interprets his expenditure of early morning time.

From a personal point of view, ignoring the "bare" essentials, my program of action involves the putting on of six items of clothes: shoes, socks, trousers, shirt, tie, and jacket. A program of action is any order in which the clothes can be put on. There are $6 \times 5 \times 4 \times 3 \times 2 \times 1 = 720$ different orderings. Many of these are not *feasible* programs as they do not meet the constraints of society (socks over shoes) or are impractical (tie on before shirt). Even after eliminating these infeasible solutions from consideration, I still have a number of feasible programs to contend with.

How is the final selection—the optimum decision—made? The dressing problem, like all those to be considered, has some measure of effectiveness—some basic objective—which enables me to compare the efficacy of the available feasible programs. If, in some fashion, I can compare the measures for each program, I can select the optimum one. For the dressing problem, I am concerned with minimizing the time it takes to get dressed. This is my measure of effectiveness—the objective function in programming terminology—with which I can compare the various feasible orderings of clothes.

Admittedly, I have not solved this problem with stopwatch accuracy, but over the years, my optimum solution has been the following ordering: socks, shirt, trousers, tie, shoes, jacket. This is my optimum solution—it minimizes the time to get dressed within the fashion constraints of society. Someone else with a different objective function—minimize the opening and closing of drawers, i.e., minimize the early morning noise—might select a different optimum solution.

The dressing problem, although not a linear-programming problem, typifies programming problems in that it has many possible solutions. If there were only one feasible solution, there would really not be any problem or any fun in solving it. There is also some objective to be optimized which enables us to select at least one of the feasible solutions to be the optimum. The finding of the feasible solutions and the determination of an optimal one represents the computational aspects of programming problems which are discussed in later sections.

As we shall concern ourselves only with linear-programming problems, it should be stressed that linear programs are a special subset of general programming problems (usually called mathematical programming) in that the mathematical description of linear programs can be written in terms of linear or straight-line relationships. For example, if one pound of coffee costs $1.00, and the seller offers no quantity discount, the total cost is directly proportional to the amount purchased; i.e., it is a straight-line relationship, as shown

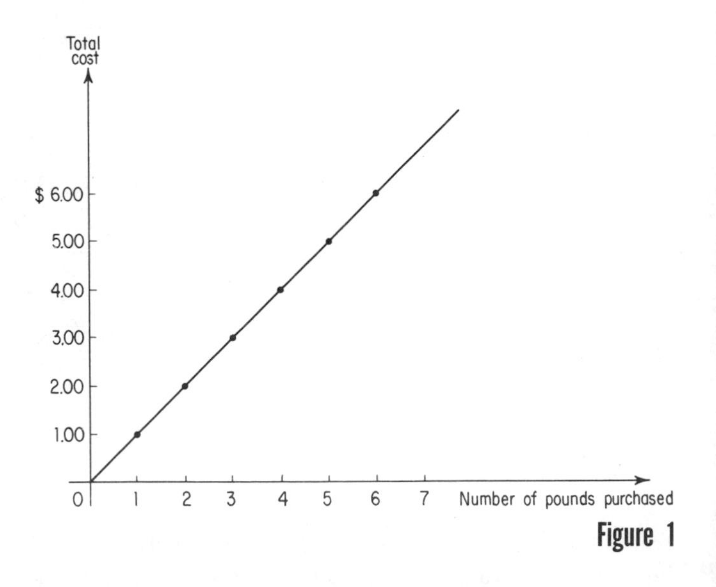

Figure 1

in Figure 1. On the other hand, if the seller allowed 10 cents off for the second pound purchased, 20 for the third, etc., up to the fifth pound, and 50 cents per pound afterwards, the cost curve would be nonlinear, as shown in Figure 2. In sum, linear-programming problems are those programming problems whose relationships—the

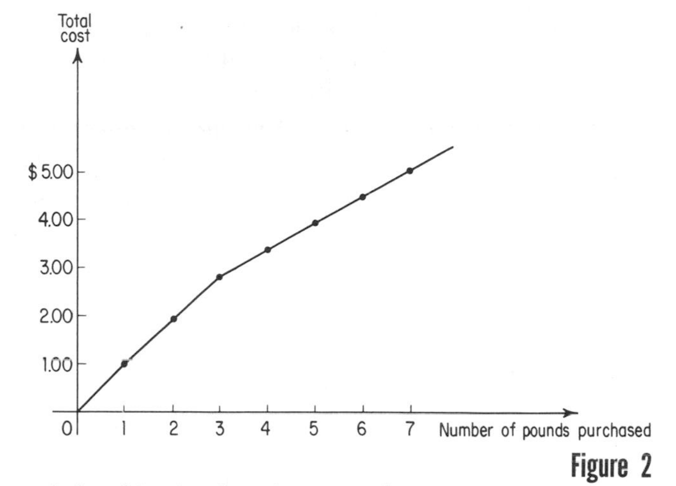

Figure 2

constraints of the problem and the objective function—can be represented mathematically as linear relationships. Although this appears to be quite restrictive, many important problems have this simplifying feature.

Programming problems, especially linear-programming problems, arise in a wide variety of settings. There are standard applications of programming techniques found in the areas of agriculture, petrochemicals, paper manufacturing, transportation, production scheduling and inventory control, military and defense, engineering, economics, to cite just a few. A bibliography of applications is given in the Appendix. We shall deal with a few such applications in order to develop the fundamentals of linear programming. But first—as a slight digression—we must discuss the place of linear programming within the scheme of modern-day scientific decision-making, i.e., operations research or management science.

OPERATIONS RESEARCH AND MODELS

When we mean to build,
We first survey the plot, then draw the model;
And when we see the figure of the house,
Then must we rate the cost of the erection.

SHAKESPEARE

Definitions of operations research (OR), like advice to the lovelorn, have always been in demand and are plentiful. They range from the detailed one of "operations research is the application of scientific techniques, tools, and methodology to problems involving the operations of a system so as to provide those in control of the system with optimum solutions to the problems," to the succinct ones of "operations research is quantitative common sense" or "research into operations." For our needs, we consider OR to be the application of scientific techniques to decision problems. What interests us most, however, is not what OR claims to be, but its methodology.

The phases of an OR project can be delineated into six overlapping and somewhat ill-defined stages.[1] For most purposes they are:

- Formulating the problem
- Developing a mathematical model to represent the system under study
- Deriving a solution from the model
- Testing the model and solution
- Establishing controls over the solution
- Putting the solution to work

Here we have introduced the concept of a mathematical model, which we shall see is central to the methodology of OR. These phases of an OR project can be viewed as accomplishing the following: For any problem we need to define the broad objectives and goals of the system—examine the environment we are working in—become familiar with the jargon, the people and things associated with the problem—determine the alternative courses of action available to the decision-maker—develop some statement, verbal or otherwise, of the problem to be investigated; translate the problem into a suitable logical or mathematical model which relates the variables of the problem by realistic constraints and a measure of effectiveness; find a solution which optimizes the measure of effectiveness, i.e., a feasible and optimum solution; compare the model's solution against reality to determine if we have actually formulated and solved the real-world problem we started with; determine when the real-world situation changes and the reflection of such changes into the mathematical model; and most important, implementation—putting the solution into operation (not just filing a report) and observing the behavior of the solution in a realistic setting. As our ability to develop precise mathematical models of operational problems is not a highly developed science—most people believe it

[1] R. L. Ackoff, The Development of Operations Research as a Science, *Journal of the Operations Research Society of America*, vol. 4, no. 3, June, 1956.

to be an art—we must be sensitive to discrepancies in the solution and feed back to the model refinements that will cause future solutions to be more realistic and accurate.

Models have been classified into three basic types. The *iconic model* looks like what it is supposed to represent. It could be an architectural model of an apartment complex, a planetarium representing the celestial sphere, or the idealization of the typical housewife as demonstrated by the designer's fashion model. The *analogue model* relates the properties of the entity being modeled with other

properties that are both descriptive and meaningful. For example, the concept of temperature is described by a graph in which a degree is equivalent to a specified unit of distance. Finally, the *symbolic model* or the *mathematical/logical model* represents a symbolic description of the process or problem under investigation. Einstein's famous equation $e = mc^2$ states, in symbols, that the energy e contained in a mass m is equal to the product of the mass and c^2, the square of the velocity of light. The mathematical model represents the translation of the statement of the problem into quantitative terms. As we shall see, the model of a linear program is a mathematical one. In programming terms, the mathematical model represents a set of relationships among the variables, resources, constraints, and objective function (measure of effectiveness). The mathematical model is central to the methodology of OR; it offers understanding of the process and problem under investigation; it provides a vehicle for the evaluation and comparison of alternative solutions; it enables us to evaluate the effects of a change of one variable on all the other variables; and finally, and somewhat mystically, it provides us with a quantitative basis to sharpen and evaluate our intuition of the process under investigation.

It should be stressed that the mathematical model is the prime distinguishing feature of OR, mathematical decision-making, or management science. Mathematical models enable us to bring some semblance of scientific methodology to areas of decision-making heretofore characterized by intuition and experience. Mathematical models abound in the areas of inventory control, allocation of resources, queuing, competitive situations, transportation, industrial processes, and many more.

The role of the mathematical model in OR and decision-making can be summed diagrammatically, as in Figure 3.[1] After the statement of the problem, which includes the choice of the all-important measure of effectiveness, the functional form of the mathematical model is determined. As this requires specifying how the variables are related with associated data, certain experiments designed to aid the structuring of the correct form must be carried out. In some instances, this experimentation could be just the opening of the accounting ledger to gather the needed information; in others, it might call for complex and expensive efforts. In any event, the results

[1] H. W. Goode, An Application of a Highspeed Computer to the Definition and Solution of the Vehicular Traffic Problem, *Journal of the Operations Research Society of America,* vol. 5, no. 6, December, 1957.

are fed back into the structure of the model as shown in Figure 3.

It is by means of the mathematical model that the problem is connected with its proposed solution. The major activity here is the devising of alternative ways of solving the problem, using the mathematical model to evaluate a proposed solution and the measure of effectiveness to choose a solution to implement. For some problems the development of a set of feasible alternative solutions is automatically accomplished by computation involving the related mathematical model. This is the case in linear programming. For others, ingenuity and innovation are a must in proposing alternatives.

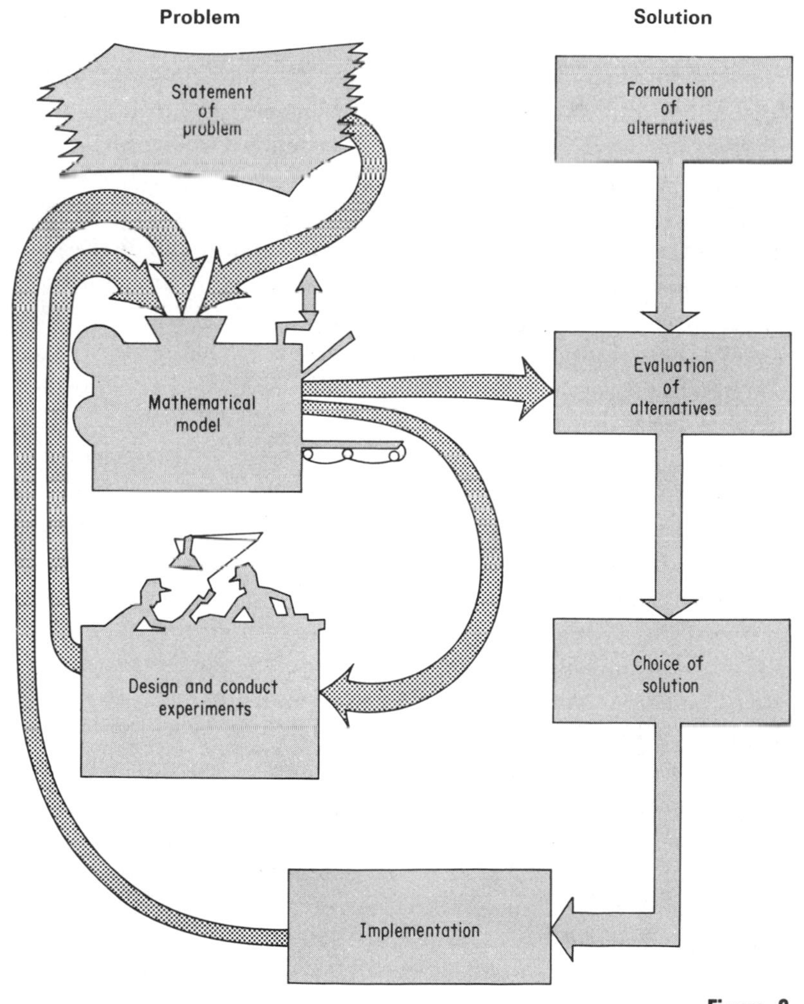

Figure 3
The role of the mathematical model.

As the implementation of a solution can affect the structure of the mathematical model, we have a feedback loop from the solution side to the problem side, and the process continues. We illustrate the above by describing a particular problem and developing its mathematical model, which in this case is a linear-programming model.

THE TRANSPORTATION PROBLEM

For my part, I travel not to go anywhere,
but to go. I travel for travel's sake.
The great affair is to move.
ROBERT LOUIS STEVENSON

A refrigerator manufacturer has two factories which supply his three retail stores. At the beginning of each month, the manufacturer receives from each store manager a list of unfilled sales which must be filled in the coming month by new production. This set of requirements represents the total number of new refrigerators that must be produced by the two factories. To simplify the discussion, we assume the manufacturer has enough resources—manpower, raw material, etc.—to fulfill the requirements, and in this instance, he does not have any over production; i.e., there is no facility for storage. The production process itself could possibly be treated via a mathematical model, but here our interests are in a different area. The manufacturer's store one, denoted by S_1, requires 10 refrigerators, S_2 requires 8, and S_3 requires 7, for a total of 25 refrigerators. He has decided to produce 11 at factory one, F_1, and the remaining 14 at F_2. The problem we wish to consider is how many refrigerators should be shipped from each factory to each store so as to minimize the total cost of transporting the refrigerators from the factories to the stores. The basic form of this problem, termed the *transportation problem,* is one of the earliest and most widely used formulations of linear programming.

We need additional information concerning transportation restrictions and costs. It is assumed that it is possible to ship any specified number of refrigerators from each factory to any store; i.e., a transportation link—rail, truck, or other mode—connects any factory to all stores. We also assume knowledge of the costs of shipping one refrigerator from a factory to a store. Here we must make a linearity assumption about these costs which is quite critical—and in some instances quite debatable. This linearity or proportionality assumption requires that if the cost of shipping one refrigerator from F_1 to S_1 is \$10, then the cost of shipping two refrigerators is \$20, for three the cost is \$30, and so on. We can argue about this assumption based on real-world experience. If it costs \$100 to hire a truck to deliver one refrigerator, the unit cost for two refrigerators (excluding handling costs) would be \$50, for three it would be \$33⅓—a nonlinear cost relationship. As another example, we note that parcel post rates (including insurance) for a \$3.80 gift from Atlantic City have the following table:

Zone	*Mileage from Atlantic City*	*Postal rate*
Local	—	\$0.69
1	150	0.93
2	150	0.93
3	300	1.20
4	600	1.32
5	1000	1.50
6	1400	1.71
7	1800	1.93
8	1800+	2.25

The graph of these rates versus mileage illustrates the nonlinearity of the postal rates, Figure 4. We see that the graph is not con-

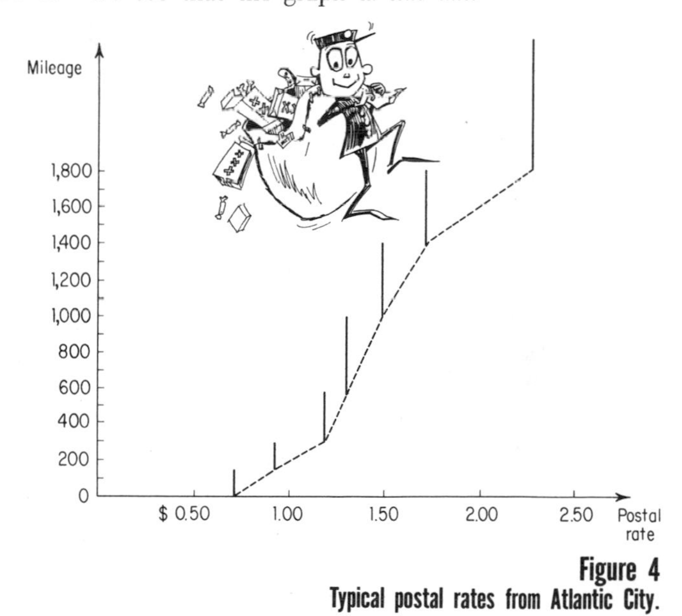

Figure 4
Typical postal rates from Atlantic City.

tinuous in that it is made up of vertical and disconnected segments. If the government allowed the postal rate to vary by the exact mileage, the nonlinear, but connected, cost graph would be the dotted line. In most transportation situations, if the cost is not linear, a good approximation can be obtained by averaging costs used by previous solutions.

For this problem then, we assume that the costs of transportation for a refrigerator between each factory and store are known and are linear. These costs are shown in the price tags of Figure 5. For

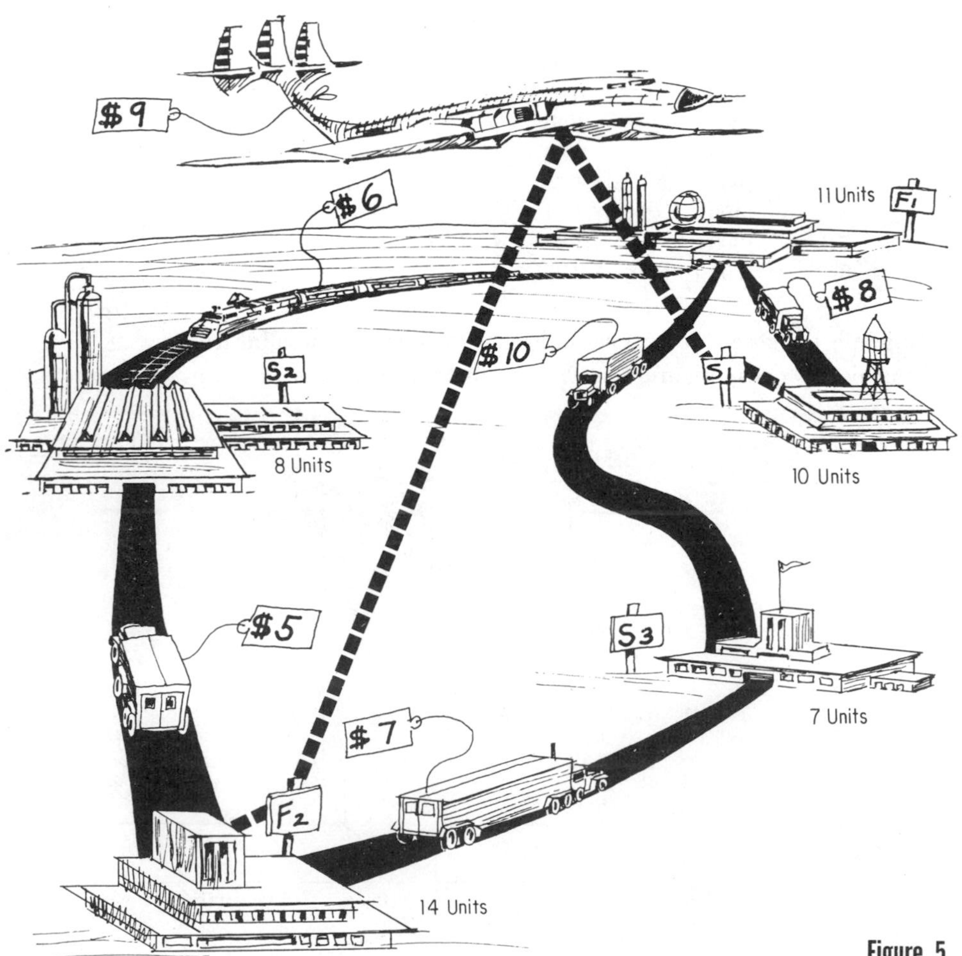

Figure 5

example, the cost of shipping a refrigerator from F_2 to S_3 is \$7. Figure 5 illustrates the ways in which the refrigerators can flow from the factories to the stores and presents all the information of the problem. In a sense, it is an iconic model. Although such a depiction does not solve the problem, it can aid us in the development of a suitable mathematical model. Transportation problems have, in general, many possible feasible solutions. We shall develop some below, along with the linear-programming model of the problem.

We are looking for a solution which meets the constraints of the problem; i.e. ship 11 units from F_1, ship 14 units from F_2, S_1 receives 10 units, S_2 receives 8 units, and S_3 receives 7 units; and, at the same time, minimizes the measure of effectiveness—the total transportation cost.

In order to proceed with the formulation of the mathematical model, we rearrange the given information in a tableau:

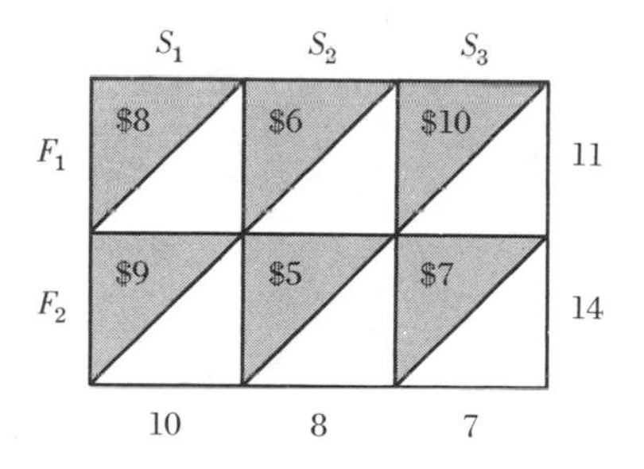

The blank triangles correspond to the unknown number of refrigerators to be shipped from the corresponding factory and store. To demonstrate that an experienced—or even inexperienced—shipping clerk (or vice-president in charge of shipping refrigerators) has no difficulty in coming up with a solution, we exhibit two possible solutions:

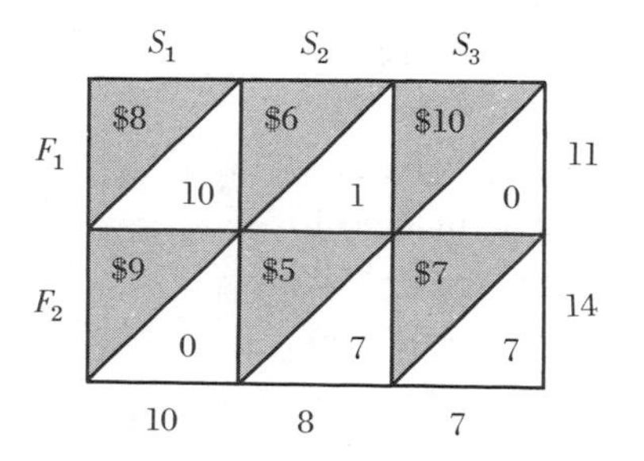

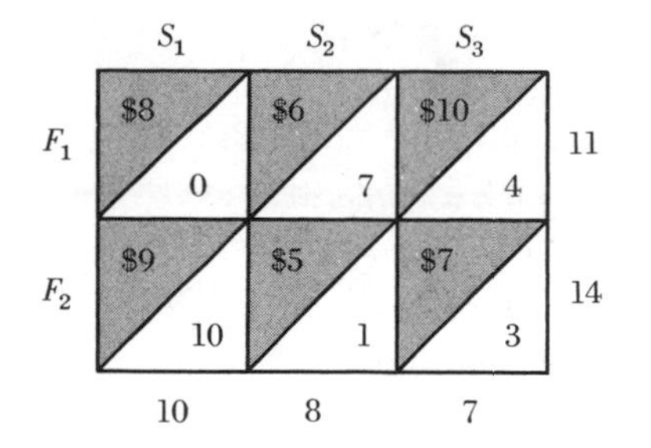

The reader will not have any difficulty writing down others. The numbers written in each box of the first solution tableau constitute a solution with the total number shipped from F_1 to each store being $10 + 1 + 0 = 11$; from F_2 we have $0 + 7 + 7 = 14$. Also, the total amount received by S_1 from both factories is $10 + 0 = 10$; S_2 receives $1 + 7 = 8$; and S_3 receives $0 + 7 = 7$. It is similar for the second solution. Assuming linearity of the cost of transportation, the total cost of the first solution is given by the expression

$$\$8 \times 10 + \$6 \times 1 + \$5 \times 7 + \$7 \times 7 = \$170$$

and the cost of the second solution is

$$\$6 \times 7 + \$10 \times 4 + \$9 \times 10 + \$5 \times 1 + \$7 \times 3 = \$198$$

For the solutions exhibited, the first has a lower cost, but the question remains as to whether there is another feasible solution which is cheaper. A shipping clerk who does not use linear-programming techniques to aid him in solving the problem must rely heavily on his experience and intuition. He does not compare exhaustively all possible solutions—neither does the linear-programming procedure. The shipping clerk does select a particular solution to implement, but in general, he cannot guarantee that he has the absolute minimum. The linear-programming approach offers an unconditional guarantee that the minimum will be determined. For the refrigerator problem, the first solution is the minimum solution.

To proceed with the development of the mathematical model of the transportation problem—and to simplify the discussion—we shall introduce some necessary mathematical shorthand. Let x_{11} be the number—the unknown number—of refrigerators to be shipped from F_1 to S_1, x_{12} the number to be shipped from F_1 to S_2, etc., and, in general, x_{ij} the unknown number of refrigerators to be shipped from factory i to store j. We enter these notations into the tableau structure as shown in the tableau:

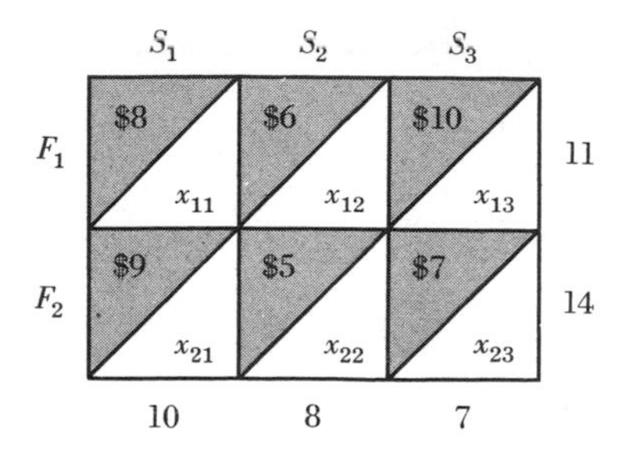

It is now a simple matter to develop the form of the mathematical model.

The total amount shipped from F_1 is 11, and the amounts shipped from F_1 are x_{11}, x_{12}, and x_{13}. Similarly, the total from F_2 is 14, and the shipments from F_2 are x_{21}, x_{22}, and x_{23}. Since the manufacturer requires a total of 25 units (10 + 8 + 7) and since he is making exactly 25 units (11 + 14), the total made at each factory must be shipped to the stores. Thus, the total amount shipped from F_1 is given by the equation

$$x_{11} + x_{12} + x_{13} = 11$$

the total shipped from F_2 is

$$x_{21} + x_{22} + x_{23} = 14$$

These sums are obtained by adding across the rows.

As each store must get exactly the amount asked for, the amounts shipped to each store—found by adding down the columns—are given by the equations

$$x_{11} + x_{21} = 10 \qquad \text{for } S_1$$
$$x_{12} + x_{22} = 8 \qquad \text{for } S_2$$

and

$$x_{13} + x_{23} = 7 \qquad \text{for } S_3$$

For any set of numbers x_{ij}, where again i represents one of the factories and j represents one of the stores, the total cost—the sum of the costs of individual shipments—is

$$\$8x_{11} + \$6x_{12} + \$10x_{13} + \$9x_{21} + \$5x_{22} + \$7x_{23}$$

These equations represent the basic constraints of the mathematical model. The only element missing is that we must limit the possible values of x_{ij} to positive values or zero. A negative x_{ij} would represent a shipment of refrigerators from a store to a factory; i.e., a source of refrigerators other than those manufactured at the factories would be introduced. We disallow this possibility by restricting $x_{11} \geq 0$ (x_{11} is greater than or equal to 0), $x_{12} \geq 0 \ldots$, $x_{23} \geq 0$, or in general notation, $x_{ij} \geq 0$. These are called the "nonnegativity restrictions" of linear programming. As we wish to determine the set of numbers x_{ij} satisfying the equations, the nonnegativity restrictions and which minimizes the total cost, we have the following mathematical model—the linear-programming model of this transportation problem:

Find the set of numbers $x_{ij} \geq 0$ which minimizes

$$\$8x_{11} + \$6x_{12} + \$10x_{13} + \$9x_{21} + \$5x_{22} + \$7x_{23}$$

subject to the constraints

$$\begin{array}{lllllll} x_{11} + x_{12} + x_{13} & & & & & & = 11 \\ & & & x_{21} + x_{22} + x_{23} & & & = 14 \\ x_{11} & & & + x_{21} & & & = 10 \\ & x_{12} & & & + x_{22} & & = 8 \\ & & x_{13} & & & + x_{23} & = 7 \end{array}$$

The first solution given above satisfies these equations, where $x_{11} = 10, x_{12} = 1, x_{13} = 0, x_{21} = 0, x_{22} = 7$, and $x_{23} = 7$, and as noted before, this solution minimizes the objective function with a value of \$170.

Each factory and store contributed an equation in terms of the variables related to the corresponding factory or store. These equations, as well as the objective function, are linear equations, since they are simple sums of the variables. The total number of variables is the product of the number of factories and the number of stores; in this case $2 \times 3 = 6$. Also, the number of equations is the sum of the number of factories and the number of stores; here it is 5. Transportation problems can become quite large, but computational procedures—algorithms—for solving rather large-sized problems are available for most electronic computers.

From a mathematical standpoint a number of interesting items should be noted about the above system of equations. First, there is one equation too many, in that any one of them is implied by the remaining ones. For example, if we drop the first equation, it can be found by adding the last three and then subtracting the second. A more important point, however, is that if we solve the problem using the standard computational procedures of linear programming, we will determine an optimum solution whose values of the variables, the x_{ij}'s, are in terms of whole numbers. It was tacitly assumed that the x_{ij}'s must be in integers—we cannot ship $3\frac{3}{4}$ refrigerators! We can prove mathematically that we will obtain an optimum integer solution for the transportation problem, given that the amounts available at the factories and required by the stores are integers. This is not the case for the general linear-programming problem. For the transportation problem, it is due to the special structure of the equations of the corresponding mathematical model.

Although we described this linear-programming model in terms of factories, stores, and refrigerators, it is quite important to recognize that we could have cast the discussion into a more general format dealing with origins (factories), destinations (stores), homogeneous units to be shipped (refrigerators), and some measure to be minimized (total transportation cost). The following discussion illustrates this point and offers another example of the linear program called the transportation model.

This transportation problem calls for the shipping of a supply of an item from a group of Air Force depots to a group of receiving stations. Each depot has a limited amount of the item. There are many routings which will supply each station with exactly the required amount of the item. The problem is to determine the routing which not only fulfills the requirements, but also minimizes some measure of the cost. For example, the objective might be to minimize one of the following: the total dollar cost, the total number of miles of the shipping schedule, or the total time the items are in transit.

Assume that Lockbourne Air Force Base (AFB) at Columbus, Ohio, has been testing a large item of equipment, weighing a ton, for the B-52. It is now desired that this equipment be tried at other bases. Five of each are required by March AFB at Riverside, California; Davis-Monthan AFB at Tucson, Arizona; and McConnell AFB at Wichita, Kansas. Pinecastle AFB at Orlando, Florida, and MacDill AFB at Tampa, Florida, each need three. To supply these 21 items, Lockbourne AFB at Columbus, Ohio, can ship 8; Oklahoma City Depot has 8; and Warner-Robins AFB at Macon, Georgia, has 5. The items of equipment are to be airlifted to their destinations. All the foregoing information, together with approximate air distances, are given in the following table:

		Required items				
City	Items available	MacDill 3	March 5	Davis-Monthan 5	McConnell 5	Pinecastle 3
Oklahoma City	8	938	1030	824	136	995
Macon	5	346	1818	1416	806	296
Columbus	8	905	1795	1590	716	854

Distance in miles

Here the objective is to *minimize the total ton-miles.* The optimum solution in terms of this objective has been computed and is given on page 21. The reader will find it instructive if he attempts to find the most efficient routing.

A first attempt at a solution might go like this: Try to fill each requirement from the nearest source. Let McConnell get its 5 items and Davis-Monthan 3 of its 5 items from Oklahoma City. The requirements for the Florida bases should be met, as far as possible, from Macon. Let Pinecastle get its 3 and MacDill get 2 of its 3 items from Macon. The rest of the requirements must be met from Columbus. This sample shipping schedule has a total of 17,792 ton-miles and is illustrated on page 21.

There are many other possible routings. The minimum solution, obtained by the methods of linear programming, has a value of 16,864 ton-miles. The computational procedure starts with any solution to the problem—such as the one described above—and in a systematic manner obtains better solutions until the optimum one has been determined. Again, linear-programming procedures guarantee that the minimum will be found.

If the reader tried to fill some or all of McConnell's needs from Oklahoma City—that is, to take advantage of the shortest distance appearing in the problem, as was attempted in the first sample solution—the resulting schedule would cost more. This emphasizes that what at first appears to be a straightforward approach to solving problems of this nature is not necessarily the best one.

The discussion of the transportation problem and its variations can be extended into a separate treatise covering an important segment of mathematical decision-making. We shall, in later sections, return to transportation-like problems. It suffices to say at this point,

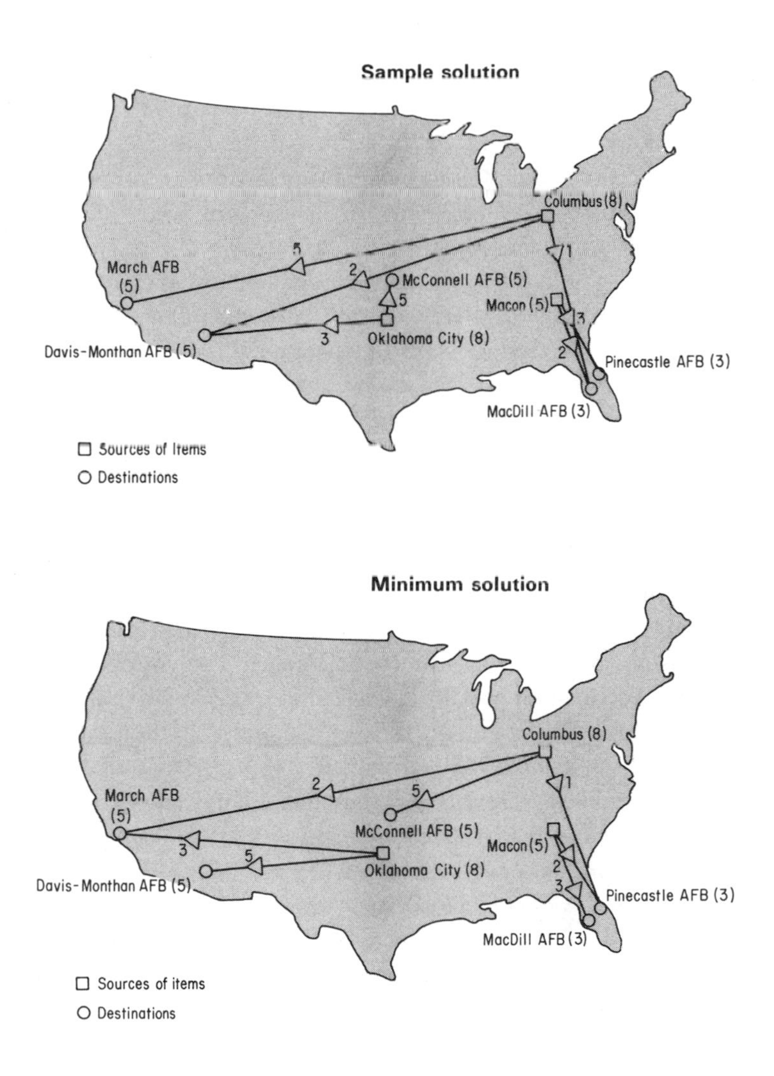

that the mathematical model of the transportation problem has proved to be quite versatile, especially when it has been coupled with the ingenuity and imagination of the new breed of scientific problem-solvers. However, one lesson I hope the reader has learned from the above discussion is that a simple, logical approach to the definition of a problem can carry us quite far along the yellow-brick road which leads to the development of the proper mathematical model. This is especially true in the field of linear programming. The complexities, which sometimes can be quite formidable, can usually be resolved without recourse to the wizard.

N.B.

In our travels through this *Guide* it is quite important for us to pause on occasion and attempt to place the material and discussion in perspective as it relates to our current and future environments. The end of this chapter marks the first such point for reflection. For the author the following thoughts are in order.

My own enthusiasm for linear programming, although shared by many others, must be tempered by the remembrance of difficulties encountered in formulating many operational problems—the difficulties of stating implied constraints, resolving conflicting objectives, determining the data needs and being unable to find the data, and the human and political interplays inherent in any situation in which it appears as if mathematics and the mathematicians are about to replace those who "really" understand the problem.

Although an important branch of applied mathematics in its own right, linear programming and its extensions are rightly subsumed within the more general field of scientific decision-making, that is operations research. In the past, OR gave the impression that its main preoccupation was the optimization of large, integrated systems like complex production-distribution facilities. In reality, however, the problems which could be resolved were subsystems of subsystems like a simplified transportation problem. The reasons for this are manyfold and are still with us—we cannot state the complex interactions in a precise mathematical model and/or we do not have the mathematical-computer techniques which can solve such problems. Thus, the simplified discussions of the linear-programming problems contained here suffer the same subsystem phenomenon as OR in general. We can solve important subclasses of problems, but the problems associated with the optimization of major systems are still with us.

What of the future? It is felt that the key to the successful operation, if not the optimization of complex systems, is the development of a computer-oriented information system which reflects the current status of the operational system. The information system must be coupled with appropriate mathematical models and solution techniques (like linear programming), along with heuristic computer procedures which enable the human element—aided by the computer—to investigate and resolve those elements of the problem which defy precise mathematical descriptions. For example, for a set of production-distribution facilities (like that of the refrigerator manufacturer), we might find that we are unable to develop a mathematical optimization model which allows us to schedule the production resources and equipment and to sequence the units to be processed. But, given the completed units, we can ship them to the proper destinations in an optimal fashion using a transportation model. Here, the information system would picture the status of the production elements and the orders to be processed. Using heuristic techniques, the manufacturer could simulate the production pattern over a range of conditions and assumptions; determine the cost for each pattern; using the transportation model, determine the minimum shipping cost for each pattern; and select the pattern which yielded the total minimum cost. This procedure will not yield the true minimum cost, as we have assumed we do not have an optimization model for the production process. Because of the speed of computers, the procedure does, however, enable the manufacturer to evaluate a number of possible production patterns—including the one his production supervisor would normally use—before actually starting production on the orders. He is thus able to optimize the overall process within the limits of current decision-making capabilities.

In this hypothesized, integrated, information-optimization-heuristic approach we see that the element of optimization is but one feature in the development of control procedures for complex problems. It is felt that linear programming will be a major building block in the construction of such procedures, just as it has been a major optimization technique for a wide range of realistic subproblems.

FORMULATION OF PROBLEMS

{We continue our travels and learn what to eat, visit the farm, and encounter some strange beings who inhabit Linear Programsville.}

THE BASIC AIM in developing a linear-programming model of an operational problem is to be able to predict what the optimum solution should be, given the initial conditions of the problem. In making this statement, I assume we have been able to capture the real-life situation—that is, the actual problem we wish to solve—by proper definition and manipulation of variables and constraints. In many instances, in fact, in most instances, our ability to portray the true, the genuine, *the problem,* is open to question. As in all other areas of human endeavor, we find that compromises, resourcefulness, and a little finagling help to obtain a better understanding of the process in question and, we hope, lead to a mathematical model which yields an improved solution that *can be put to work.* A model which claims to portray the production capacity of a manufacturing plant can only do so within certain limits. The predicted production, which is used to plan transportation, storage, and related needs, is based on assumed or measured processing rates, availability of manpower, resources, etc. What actually happens—a slowdown here, a shortage there—molds and forces the events which render the true production. If the model is a "reasonable" mathematical representation of the real problem, the plans based on the predicted production have not led us too far astray and, in fact, have allowed us to plan the operation in a more efficient and, thus, a more profitable manner.

In developing the formulation of problems in terms of a linear-programming model, we must guard against being accused of having a tool for a job and, if the tool does not fit, reshaping the job to fit the tool. This *caveat* does not and should not rule out our being allowed to make proper simplifying assumptions in light of the desire to develop models which capture the essentials of the problem. Our mathematical models must yield answers that can be understood by the people responsible for the process being studied. These people—sometimes loosely termed "the decision-makers"—must be able to put the results to work. These results must cause the operation to improve in terms of the agreed-upon measure of effectiveness.

How do we go about formulating the mathematical model of a linear-programming problem? What are the danger spots? How can we get results that work? The answers to these and related questions cannot be delineated in clear terms. At best, I can talk around the answers and impart their meaning by the discussions and illustrations that follow.

As noted earlier, although I first introduced the transportation problem as one in which we shipped goods from factories to stores, the general transportation problem can involve generic origins and destinations. I must emphasize this point. I shall illustrate a number of linear-programming problems by imbedding the problem in a specific environment. The reader—who by this time must have recognized the great burden I have placed upon him—should not be of such a narrow mind as to believe that the resulting mathematical model does not have a use beyond the given setting. The following fable illustrates the point.

In the early days of linear programming, circa 1953, one of the few formal publications in the field described the essentials of problem formulation in terms of determining an optimal product mix. A processor of nuts wished to mix three grades of nuts, each mix consisting of cashew nuts, hazel nuts, and peanuts, in such a fashion as to meet certain specifications. For example, one mixture had to contain not less than 50 percent cashews and no more than 25 percent peanuts. The processor wished to combine his available resources of nuts, subject to the capacity restrictions of his manufacturing plant, so as to maximize his profits. The problem was explained in complete detail, the mathematical model developed, and a numerical example was solved. The example and the publication had a wide audience.

Included in this audience was one of the pioneering operations-research consultants who worked in Detroit. The consultant was approached by the production manager of a large automobile manufacturing company who inquired how he could learn about this new technique called linear programming. The OR man briefed him on the subject and recommended the nut-problem publication for further information.

A few weeks later he called the production manager to inquire about the progress of the self-study course. The manager was perplexed. He chastised the OR consultant and mentioned something about wasting his time. Undaunted, the consultant pressed on in order to determine the trouble. Finally it came out. The manager had read and reread all about the nut problem. He felt that he might now be able to become a big production man in a nut factory—but he made automobiles. The manager was unable to transfer the concepts of the product-mix, resource-allocation nut problem to his own environment. He could not "relate." I am sure today's readers are more astute. At least they have been forewarned.

To illustrate the versatility and adaptability of the linear-programming model, I shall next describe a number of the now-classical problems found in the linear-programming literature. Mathematically, these problems are all of the same form and can be solved by the standard computational method of linear programming, *the simplex method.* This computational procedure is grounded in advanced mathematical considerations, and a proper treatment would be beyond the scope and aims of this book.[1] My purpose here is to have the reader become adroit in recognizing the essential features of linear programming and not to become proficient in solving such problems. Thus, computational discussions will be rudimentary in nature and are relegated to the next chapter and the Appendix.

THE DIET PROBLEM

The proof of the pudding is in the eating.
CERVANTES

Anyone attempting to live within a budget has a number of alternative ways to allocate his limited funds. A budget-minded housewife has

[1] The reader who wishes to pursue the mathematical road is referred to the Appendix and the author's textbook, *Linear Programming: Methods and Applications,* 3d ed., McGraw-Hill Book Company, New York, 1969.

to set aside so much for rent, clothes, food, entertainment, transportation, and so on. The fixed costs like rent are easy to allocate, while the split of money between food and entertainment is usually made based on past experience, with occasional spur-of-the-moment fluctuations. The specific allocations, however, are made based on the housewife's measure of effectiveness, evaluated with respect to the expenditure of her funds in all the budget areas. In these types of decision problems, it is rather difficult to optimize or even to determine an overall measure of effectiveness, and thus, one attempts to suboptimize—break up the problem into tractable subproblems, each subproblem consisting of an applicable measure of effectiveness and associated constraints. For example, let us consider the subproblem faced by the housewife as she attempts to feed her family. In fact, we shall greatly simplify the problem—as shall be our approach to such matters—by only considering her plans for getting her children to eat the proper breakfast.

Forgetting the budget constraint for the moment, the harried housewife wishes to feed her children a breakfast menu which contains a specified level of nourishment. After consultation with her vitamin/calorie counter, she decides that her children should obtain at least 1 milligram of thiamine, 5 milligrams of niacin, and 400 calories from their breakfast foods. The children have a choice of eating the latest in dry cereals—Krunchies, the old standby Crispies, or, as children are wont to do, a mixture of the two. The fine print on the side of each cereal box contains, among other things, the fact that 1 ounce of Krunchies contains 0.10 milligram of thiamine, 1 milligram of niacin, and 110 calories; while 1 ounce of Crispies contains 0.25 milligram of thiamine, 0.25 milligram of niacin, and 120 calories. We see that it is quite easy to find menus, i.e., solutions, to this problem. The desired level of nutrients can be reached or exceeded by eating 10 ounces of Krunchies or 20 ounces of Crispies. But what of the cost of such a diet, what of the budget? The objective of the housewife is to plan a breakfast menu such that her children obtain at least the specified nutrients at the minimum cost. If Krunchies cost 3.8 cents per ounce and Crispies 4.2 cents per ounce, we see that the two suggested solutions would cost 38 cents and 84 cents, respectively. The first solution, to eat 10 ounces of Krunchies, contains 1 milligram of thiamine, 10 milligrams of niacin, and 1100 calories; while the second solution, to eat 20 ounces of Crispies, contains 5 milligrams of thiamine, 5 milligrams of niacin, and 2400 calories. The first solution gives exactly the right amount of thiamine and too much of the other nutrients, and the second solution contains exactly the right amount of niacin and too much of the others. If the children wish to eat only Krunchies, the housewife cannot lower the amount of Krunchies below 10 ounces, as a lower quantity would not contain at least 1 milligram of thiamine. Similarly, if the children wish to eat only Crispies, they must eat 20 ounces, for otherwise they would not get enough niacin. The only alternative solution left is a mixture of the two cereals—the housewife must determine if there exists a combination of the two which contains enough of the nutrients and is cheaper than the 10-ounce Krunchies solution. In fact, she wants the cheapest mixture which satisfies the nutrient constraints. How can she determine the minimum cost diet? Linear programming to the rescue!

During the above discussion we have tacitly assumed certain linear relationships with respect to combinations of the two cereals. First, we assumed that the amount of nutrients ingested is proportional to the amount of food eaten; e.g., 1 ounce of Krunchies contains 110 calories, while 2 ounces contains 220 calories, etc. Also, the amount of nutrient contained in one food is additive to the amount contained in the other; e.g., 1 ounce of Krunchies and 1 ounce of Crispies contain 110 + 120 or 230 calories. The constraints of the problem state that the children must take in at least so much of a nutrient, and if a solution causes them to have an excess of a nutrient, this is acceptable. The 10 ounces of Krunchies solution contained 1100 calories, while only 400 are required. To formulate the linear-programming model let us rearrange the data of the problem in the following tableau. We denote the amount of Krunchies to be eaten by K and the amount of Crispies by C.

Nutrients	Amount of each nutrient in 1 ounce of Krunchies (K)	Amount of each nutrient in 1 ounce of Crispies (C)	Requirement of each nutrient
Thiamine	0.10	0.25	1
Niacin	1.00	0.25	5
Calories	110.00	120.00	400
Cost	3.8 cents per ounce	4.2 cents per ounce	

To construct the constraints of the problem we note that the total amount of each nutrient in the left-hand columns must be greater than or equal to ($\geq$) the numerical amount shown in the right-hand column. Thus, for each nutrient we have a constraint. For thiamine, K ounces of Krunchies contain $0.10K$ milligram, and C ounces of Crispies contain $0.25C$ milligram. A solution to the problem must have

$$0.10K + 0.25C \geq 1.$$

Similarly, for niacin, we must have

$$1.00K + 0.25C \geq 5$$

and for calories,

$$110.00K + 120.00C \geq 400$$

The amount of each food eaten must be zero or a positive amount; i.e., $K \geq 0$ and $C \geq 0$. Finally, the total cost of the breakfast menu is given by

$$3.8K + 4.2C$$

Putting the above together, the housewife's breakfast-menu problem is to find values of K and C which minimizes the total cost

$$3.8K + 4.2C$$

subject to

$$\begin{aligned} 0.10K + \quad 0.25C &\geq 1 \\ 1.00K + \quad 0.25C &\geq 5 \\ 110.00K + 120.00C &\geq 400 \\ K \qquad\qquad\quad &\geq 0 \\ C &\geq 0 \end{aligned}$$

The reader can readily check that our two previous menus $K = 10$, $C = 0$ and $K = 0$, $C = 20$ satisfy the above inequalities. The minimum solution is when $K = 4\frac{4}{9}$ ounces and $C = 2\frac{2}{9}$ ounces, with a total cost of $26\frac{2}{9}$ cents. This solution will give the children 1 milligram of thiamine and 5 milligrams of niacin, which are the exact requirements, and $755\frac{5}{9}$ calories, an overage of $355\frac{5}{9}$ calories.[1]

This simple diet or menu-planning model can be extended to include the problem of determining a three-meal diet such that minimum daily requirements for all the basic nutrients are met in a manner which minimizes the total cost of the foods used in making the meals. This type of diet problem was first formulated in the early 1940s—before the discovery of the mathematics and solution procedures of linear programming. At that time, an economist, George J. Stigler, formulated a 77-food, nutrient-diet problem, using 1939 costs of the foods. Stigler's approach to solving his problem was by trial and error. By diligent analysis and keen insight he determined the types and amounts of each food which would satisfy the daily minimum nutrient requirements for a very low-cost, but not minimum, diet. His solution called for the use of only *five* foods at a total yearly cost of $39.93. The foods were wheat flour, evaporated milk, cabbage, spinach, and dried navy beans. The true mini-

[1] How the solution to this diet problem was determined is shown in Chapter 3.

mum cost diet obtained by linear-programming methods called for nine foods—wheat flour, corn meal, evaporated milk, peanut butter, lard, beef liver, cabbage, potatoes, and spinach, with a slightly better yearly cost of $39.67. Such diets, although quite inexpensive, are certainly unpalatable over any period of time, and the selection of foods would do justice to the chief dietician of a slave labor camp.

As Stigler points out, "No one recommends these diets [i.e., true minimum cost diets] to anyone, let alone everyone." He also cites a low-cost diet for 1939 that was constructed by a dietician and cost $115. The difference in cost was attributed to the dietician's concern with the requirements of palatability, variety of diet, and prestige value of certain foods. More recent attempts at constructing diets for human beings using linear programming have met with acceptance due to the problem formulator being able to express via linear constraints the dietician's concern for flavor, taste, and variety. Such menu planning is now being done for large institutions, with a reasonable cost savings over the standard dietician's approach. But, even if the success of the diet problem for people has been limited, the diet problem for chickens, cattle, and pigs has been a notable linear-programming example.

Diet problems, or, more generally, blending problems, arise in a number of manufacturing activities. We have problems of minimum cost feed mixtures for farm animals or the mixing of various elements, e.g., chemicals or fertilizers, to meet minimum requirements at least cost. The mathematical model of these problems is an extension of the simple problem faced by our housewife. Even for animals we have such restrictions as palatability, but they can be taken care of in a straightforward manner. Although the mathematics of such applications are standard, the ability to determine the full set of proper restrictions depends on the analyst's understanding of the problem area. In one attempt to develop a feed mix for cows, a restriction limiting the total amount of molasses was included to insure that the resultant feed could be processed by the manufacturing equipment which pressed the feed into pellets. It turns out that molasses is cheap and reasonably high in calcium and protein, and the cows love it. Thus, the optimal solution included as much molasses as possible, but the optimum feed was rejected by the manufacturer. Although he could manufacture it in pellet form, the molasses content would make the manure too soft to use as fertilizer.

This rather earthy example illustrates how an analyst must go about developing a mathematical model which produces solutions that can be put to work. It is an evolutionary process and calls for the close interactions of the people with the problem and the man with the model.

THE CATERER PROBLEM

While we wait for the napkin, the soup gets cold,
While the bonnet is trimming, the face gets old,
When we've matched our buttons, the pattern is sold,
And everything comes too late—too late

FITZHUGH LUDLOW

As it must happen to all good things, a management science consultant was called in to "unmadden" the Mad Hatter's tea parties. For all these many years, the Hatter, his friends, and guests have been going around and around the table encountering dirty place settings. The table was by now becoming rather unsightly.

"If you plan to entertain each day, you must plan ahead," expounded the consultant to the Hatter.

"But I have no time to plan," complained the Hatter, "as I must entertain my friends with poems and riddles and giving them more or less tea, as the case may be."

"Leave it all to me," said the consultant. "I have observed your affairs for some time now, and with a little cooperation on your part and my model, we shall be running bigger and better and cheaper parties before you know it."

"My parties have been everyone's model for years. I would like to learn about yours. You're hired," said the Hatter.

The guests at the table all moved up one and the consultant joined them, pen and contract in hand. He and the Hatter signed, and it was duly witnessed by the March Hare. The contract called for the consultant to study the problem in depth and to submit his report and recommendations in thirty days. The report follows.

SMC

AN ANALYTICAL ANALYSIS OF INTERACTIVE ACTIVITIES AS RELATED TO THE ECONOMICS OF FUNCTIONAL GATHERINGS

A Preliminary Linear Programming Model of a Tea-party Subsystem

by

Super Management Consultants

Introduction

A brief study of tea-party operations as manifested by the present management, Mad Hatter, Inc., led us to the immediate and conclusive conclusion that things were in a terrible state of affairs. Any attempt to bring some semblance of analytical methodology to the present operations would meet with disaster. Thus, we recommend that tea-party operations be stopped as soon as present invitations are honored and a program of action, as outlined below, be initiated as soon as possible. Although the suggested program considers only one aspect of the total tea-party operations, it is felt that if we can accomplish a cost-effective program for this pivotal part of the operations, we can then push on to full tea-party service with a high probability of success. More specifically then, we recommend that all of the present stock of dirty napkins be destroyed and a purchase-laundering programming for new napkins be instituted.[1] We have attacked this element of the total problem using the powerful tools of modern mathematical decision-making. Our approach follows.

---

[1]Controlled experiments showed that the napkins were in such a deplorable condition they could not be cleaned.

The Mathematical Model

Based on our extensive experience in mathematical modeling, we were able to structure within the total tea-party system a critical subsystem which lends itself to a complete mathematical analysis. This analysis is based on the well-known caterer problem, in which a caterer wishes to determine how many napkins he must purchase and how many dirty ones he must send to the laundry in order to have enough for his customers. He wants to achieve a proper balance between purchases and laundry so as to minimize the total cost of the napkin subsystem. Our plan is to adapt this linear-programming model to the tea-party problem and, once we have optimized the flow of napkins, to extend our analysis to the other elements of tea-party operations. We feel confident that within time we shall optimize the flow of food, the flow of guests, and, finally, the flow of tea.

Analysis

To illustrate the application of the caterer's problem model to the operations under consideration, we undertook a data-collection phase of our project and determined for a typical week that the following number of persons attend your tea parties:

1. Monday--5
2. Tuesday--6
3. Wednesday--7

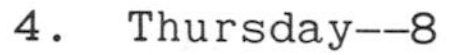

4. Thursday--8
5. Friday--7
6. Saturday--9
7. Sunday--10

Any purchases of new napkins can be made on the desired day and delivered in time at a cost of 25 cents per napkin. (Delivery is free.) There are two laundries in the area which have been tested and approved. The King's Laundry can clean a napkin in two days and charges 15 cents per napkin; while the Queen's Laundry takes three days and charges 10 cents per napkin. Assuming we have burned all the old napkins and have no new ones on hand, we next construct the linear-programming model for your data.

First, let us state our notation. We let n_1, n_2, n_3, n_4, n_5, n_6, n_7 be the number of new napkins purchased to be used on the corresponding day of tho week. Similarly, we let k_1, k_2, k_3, k_4, k_5, k_6, k_7 and q_1, q_2, q_3, q_4, q_5, q_6, q_7, be the number of napkins sent to the King's and Queen's laundries, respectively. Finally, let d_1, d_2, d_3, d_4, d_5, d_6, d_7, be the number of dirty napkins which are not sent to a laundry on the corresponding day. As we want to minimize the total cost of maintaining the proper inventory of clean napkins, it does not pay to buy more napkins than needed for the current day, or to send out a napkin to be laundered unless it will be used at a future time.

For the first day of operation, i. e., on Monday, we must buy exactly the number needed. As we expect five guests, we conclude that

$$n_1 = 5.$$

At the end of Monday's tea party, we have a choice of sending all or some of the five dirty napkins to the fast King's Laundry, to the slower Queen's Laundry, or letting them stay dirty in the laundry room. What happens to these five napkins can be represented by the equation

$$k_1 + q_1 + d_1 = 5$$

The total cost for the first day's operation is $(25n_1 + 15k_1 + 10q_1)$ cents. Of course, our problem is to determine exactly what numerical values to give to the variables of the problem--so far we have n_1, k_1, q_1, and d_1 as variables, with n_1 equal to exactly 5. Once we have all the equations of our problem, the computational procedures of linear programming can be put to work to find the minimum cost solution. We next develop the rest of the model's constraints, remembering that laundered napkins will be back in the system in 2 or 3 days, depending on the service. The k_1 napkins sent on Monday will return in time to be used for Wednesday's tea party, while the q_1 napkins will be ready for Thursday. Also, the d_1 napkins not sent out on Monday can be sent out the next day.

For Tuesday's tea party, we again must buy the required number of napkins, i.e.

$$n_2 = 6$$

After they have been used, the actions taken to dispose of these six napkins, and the dirty pile of d_1 napkins, is given by the equation

$$k_2 + q_2 + d_2 = 6 + d_1$$

This last equation states that we now have a dirty stockpile of d_1 from Monday and six more from Tuesday that can be either laundered or kept in a dirty pile. The cost of Tuesday's operation is

$$(25n_2 + 15k_2 + 10q_2) \text{ cents}$$

For Wednesday's affair, we need seven napkins. This is the first day in which laundered napkins from the King's fast service can be put into use again. Hence, the required seven clean napkins can be obtained either by purchases and/or from the amount sent to the fast laundry on the first day. We then have

$$n_3 + k_1 = 7$$

Also, as before

$$k_3 + q_3 + d_3 = 7 + d_2$$

and Wednesday's cost is

$$(25n_3 + 15k_3 + 10q_3) \text{ cents}$$

The eight clean napkins needed for Thursday can be new ones or laundered ones which have returned from Tuesday's dirty shipment to the King's Laundry or Monday's dirty load returned from the Queen's Laundry. This is represented by

$$n_4 + k_2 + q_1 = 8$$

with

$$k_4 + q_4 + d_4 = 8 + d_3$$

and a Thursday cost of

$$(25n_4 + 15k_4 + 10q_4) \text{ cents}$$

We can now proceed to write the remaining equations of our model in a straightforward manner. We assume for discussion purposes that we are interested in the efficient running of only one week of tea parties, and hence, no laundry shipments will be made unless they can be returned to be used on Sunday.

To continue, we have for Friday the equations,

$$n_5 + k_3 + q_2 = 7$$

$$k_5 + d_5 = 7 + d_4$$

with a cost of

$$(25n_5 + 15k_5) \text{ cents;}$$

and for Saturday,

$$n_6 + k_4 + q_3 = 9$$

$$d_6 = 9 + d_5$$

with a cost of

$$(25n_6) \text{ cents;}$$

and for Sunday,

$$n_7 + k_5 + q_4 = 10$$

$$d_7 = 10 + d_6$$

with a cost of

$$25n_7$$

Putting the above equations together, we see that our problem is to find values of n, k, q, and d (these values are either positive or zero, i.e., nonnegative) which minimize the cost function 25 $(n_1 + n_2 + n_3 + n_4 + n_5 + n_6 + n_7) + 15\ (k_1 + k_2 + k_3 + k_4 + k_5) + 10\ (q_1 + q_2 + q_3 + q_4)$ and satisfy the linear equations

$$
\begin{aligned}
n_1 \qquad\qquad\quad &= 5 \\
n_2 \qquad\qquad\quad &= 6 \\
n_3 + k_1 \qquad\quad &= 7 \\
n_4 + k_2 + q_1 &= 8 \\
n_5 + k_3 + q_2 &= 7 \\
n_6 + k_4 + q_3 &= 9 \\
n_7 + k_5 + q_4 &= 10 \\
k_1 + q_1 + d_1 &= 5 \\
k_2 + q_2 + d_2 &= 6 + d_1 \\
k_3 + q_3 + d_3 &= 7 + d_2 \\
k_4 + q_4 + d_4 &= 8 + d_3 \\
k_5 \qquad + d_5 &= 7 + d_4 \\
d_6 &= 9 + d_5 \\
d_7 &= 10 + d_6
\end{aligned}
$$

Optimum Solution

Our calculations show that a clean napkin will be available for each guest if the purchases and laundry shipments indicated below are followed. The total cost would be \$8.80, with only 21 new napkins purchased to service the 52 guests expected during a week.

Purchases and Shipments

for a Typical Week of Tea Parties

	$n_1 = 5$			$d_1 = 0$
	$n_2 = 6$			$d_2 = 0$
	$n_3 = 7$	$k_1 = 0$		$d_3 = 0$
	$n_4 = 3$	$k_2 = 0$	$q_1 = 5$	$d_4 = 0$
	$n_5 = 0$	$k_3 = 1$	$q_2 = 6$	$d_5 = 2$
	$n_6 = 0$	$k_4 = 3$	$q_3 = 6$	$d_6 = 9$
	$n_7 = 0$	$k_5 = 5$	$q_4 = 5$	$d_7 = 10$
Totals:	21	9	22	21

Total cost: $21 \times .25 + 9 \times .15 + 22 \times .10 = \$8.80$

We feel confident that a mathematical approach to the running of tea parties can bring about great savings. We trust that the above analysis will be implemented and extended throughout the operations of Mad Hatter, Inc.

Historical Footnote

The caterer's problem first appeared in the literature in the guise of a military application. This thinly veiled attempt to hide the true significance and power of this model was quickly surmounted. For completeness, we note the original statement of the problem. Instead of a caterer, a military commander needs to supply aircraft engines (napkins) based on specified requirements (number of guests per day). He has a choice of buying new ones or scheduling the overhaul of repairable engines to make them available again to meet the requirements. The overhaul can be accomplished in an expedited manner (fast service) or in the normal, less costly fashion (slow service). Of course, a new engine costs more than either of the overhaul procedures.

We cite this military application to demonstrate that the adaptability of many of the linear-programming models is a function of the ingenuity of the analyst. One can, for example, show that the caterer's problem is really a transportation problem in disguise.

THE TRIM PROBLEM

This was the most unkindest cut of all.
SHAKESPEARE

As part of the course requirements of the graduate school's class in Advanced Topics in Operations Research—OR 41.519/Fall Semester, each student must visit an industrial or operationally oriented organization and develop a case study for some aspect of the company's activities. In the past, ground-breaking term papers submitted included "Mathematical Decision-making as Applied to Garbage Collecting," "Queueing Discipline within a Supermarket," and "Traffic Control within a Hydrous Environment—A Case Study of Traffic Flow in Venice, Italy."

This year's class formed the usual two- or three-man teams; each group was assigned a company and instructed to delimit a problem area. This section describes the work of one of these teams.

The company selected for our team was a highly industrialized chemical and chemical-goods manufacturer. Such an organization has the full range of scheduling, inventory, transportation, and related operational problems. After a tour of the various plants, our team decided to study the operations of the company's cellophane-manufacturing activity. It was fascinating to watch how the finished product came to be.

Viscose, which originates from the reaction of wood pulp, sodium hydroxide, and other chemicals, is forced through a long narrow slit into a bath containing sulfuric acid. The viscose is immediately converted into cellulose in sheet form. The film then passes through a series of baths where it is purified, washed, dried over heated rollers, and finally wound—like a carpet—into large rolls of finished cellophane.

These rolls of cellophane are then stacked in a warehouse waiting to be sliced into rolls of smaller width, depending on the orders to be filled. Each large roll is 5 feet wide. When a roll is to be cut into smaller-sized rolls, it is transported to the cutting room and mounted on the cutting machine. This device unrolls the 5-foot sheet of cellophane, but as it does so, the cellophane is sliced by prepositioned blades, and the smaller widths are then rolled at the other end of the machine. The blades are set to yield a number of smaller rolls for customer orders, and if possible, any leftover roll would have a width which is saleable. For example, a 5-foot roll (60 inches) could be cut into three 15-inch rolls, a 10-inch roll, and a 5-inch roll. But, if no one ever orders a 5-inch roll, then this small roll has to be turned into scrap material or destroyed; it is called the trim loss for that setting of the cutting blades. Each setting of the blades (here we require four blades to cut the 5-foot roll into the five smaller-width rolls) yields rolls which can be marketed or are trim loss.

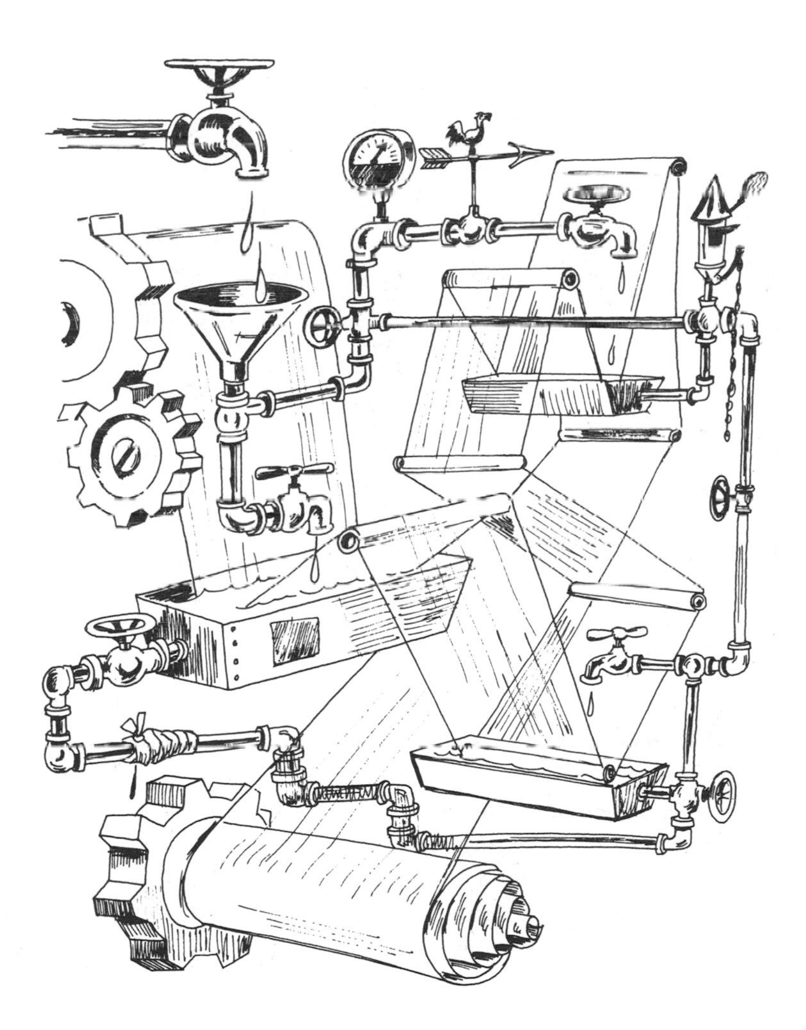

As the week's orders for the different-width rolls are accumulated, the foreman of the cutting room attempts to group them together so he can have the 5-foot rolls cut into widths to satisfy the orders and at the same time minimize the trim loss. An experienced foreman can juggle and combine the orders so as to fill all the requirements for the different sizes and does a pretty good job in keeping the trim loss rather low.

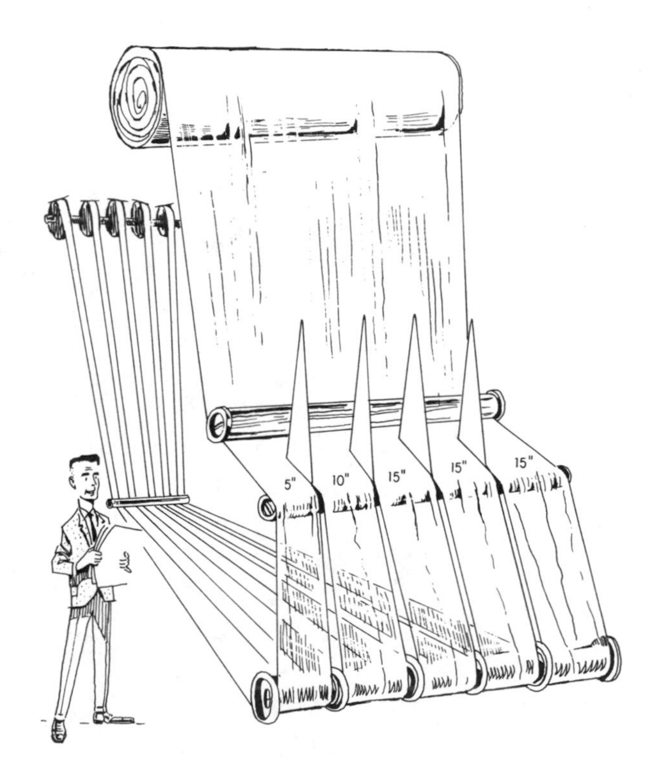

During their tour of the plant the members of the study team noted piles of trim loss being shredded and packed off. Could they cut down on this apparently wasteful and costly part of the operation? They were quick to note that the real decision-making activity was imbedded in the experience of the foreman, and then attempted to quantify his heuristic approach to optimization. Based on work originally done at the Abitibi Power and Paper Company (say it fast three times) of Canada, the team proposed to try out a linear-programming model which yielded substantial savings for the paper industry. They decided to illustrate the model by comparing the foreman's decisions with those recommended by the solution of the linear program for the same week's data. In situations like this where the newfangled mathematics is pitted against human experience, it is psychologically and strategically important to obtain the cooperation and interaction of the human element. Although novices at this game, the members of our team were astute enough to make friends of the foreman and to involve him in their experiment—or so they thought.

As supplied by the foreman, the test week's orders for rolls were as follows:

Widths ordered	*No. of rolls ordered*
28″	30
20″	60
15″	48

He was required to cut at least thirty 28-inch-wide rolls, sixty 20-inch-wide rolls, and forty-eight 15-inch-wide rolls from the supply of standard 60-inch-wide rolls. It is assumed that there are enough of the large rolls available to yield a week's orders for the smaller rolls. For the type of cellophane processed that week, any leftover roll less than 15 inches wide was to be considered a trim loss.

Key to the formulation of the corresponding linear model is the definition of the appropriate variables. Each setting of the cutting blades yields a set of smaller rolls, and the problem then becomes the determination of how the blades should be positioned and, for a given setting of the blades, how many large rolls should be cut. For example, to obtain rolls 28-inches wide, two blades can be used to cut the 60-inch-wide roll into three pieces—two rolls 28-inches wide and one roll 4-inches wide. The former rolls can be used to fill the orders, while the latter roll is trim loss. Thus, the first thing to be done is to determine which setting of the blades yields rolls that can be used. Each distinct setting is a variable of the problem, and the value of each variable represents how many 60-inch rolls should be cut at the corresponding setting of the blades. We let x_1 be the number of times we cut a 60-inch roll into two 28-inch rolls, with a 4-inch trim loss, and let other appropriate blade settings be variables as noted in the following table:

Widths ordered, in.	x_1	x_2	x_3	x_4	x_5	x_6	x_7	No. of rolls ordered
28	2	1	1	0	0	0	0	30
20	0	1	0	3	2	1	0	60
15	0	0	2	0	1	2	4	48
Trim loss	4	12	2	0	5	10	0	

Variable x_3 represents the number of times a 60-inch roll is cut into one 28-inch roll and two 15-inch rolls, with a trim loss of 2 inches. The other variables are similarly defined. Each variable is non-negative (≥ 0)—we cut rolls at that blade setting or we do not. The mathematical model is now rather easy to set down.

In setting up the equations our team determined that to give greater flexibility or freedom of action in the decision process, the manufacturer allows for more rolls of the desired widths to be cut than he has orders for. Any overage can be stored and used to help fill the next week's orders. If this were not the case, then we would have to cut exactly thirty 28-inch rolls, sixty 20-inch rolls, and forty-eight 15-inch rolls. Hence, our constraints can allow us to make more than the ordered amounts and will be represented as inequalities instead of equalities.

After constructing the above tableau, our team was able to write the three major constraints of the linear-programming model, one constraint for each width. For the 28-inch rolls we have

$$2x_1 + x_2 + x_3 \geq 30$$

for the 20-inch rolls,

$$x_2 + 3x_4 + 2x_5 + x_6 \geq 60$$

and for the 15-inch rolls,

$$2x_3 + x_5 + 2x_6 + 4x_7 \geq 48$$

Also, all the x's must be greater than or equal to zero (≥ 0). Finally, the total trim loss is measured by

$$4x_1 + 12x_2 + 2x_3 + 0x_4 + 5x_5 + 10x_6 + 0x_7$$

The team had to find the set of x's which satisfied the inequalities and minimized the total trim-loss measure.

One solution, not necessarily the minimum solution, should be the one used by the foreman. Other solutions are readily available. If $x_1 = 15$, $x_4 = 20$, and $x_7 = 12$, and all other x's zero, we see that the constraints are satisfied as equalities with a total trim loss of $4x_1 + 0x_4 + 0x_7 = 4(15) = 60$ inches. This solution states that we should cut 15 large rolls using the blade positions of x_1 (cut two 28-inch rolls and one 4-inch roll), cut 20 large rolls with the cutting pattern of x_4 (cut three 20-inch rolls), and cut 12 large rolls as given by the x_7 (cut four 15-inch rolls)—this yields a loss of only 60 inches. Are there any other solutions with a smaller trim loss?

We see if $x_3 = 30$, $x_4 = 20$, and $x_7 = 12$, we again satisfy the constraints (we have more 15-inch rolls than needed), and we also have a 60-inch trim loss. We leave it to the reader to demonstrate that either of these solutions is the minimum-trim solution.

With their solution in hand, our team reviewed the approach with the foreman and attempted to compare his decisions with the model's results. The foreman noted that such a comparison was impossible and that the team abstracted the real-life problem to fit their model —and he proceeded to show them where they went wrong.

The decision process of how to cut a roll could not be made independent of the physical characteristics of an individual roll. Where in the paper industry a roll is made without imperfections, the cellophane rolling process in use at this plant rolled the 5-foot roll in an erratic manner. Most rolls looked like a rolled carpet with one end pushed in and the other telescoping out. It was rare that a full 5-foot roll was available. The ends had to be lopped off and the resultant shorter roll used to fill the orders. A good portion of the trim loss being shredded and packed was due to the manufacturing process instead of the cutting process. Also, because of the difficulty in keeping the proper tension on the cellophane as it was rolled from the vats, bad spots appeared in some rolls and had to be cut away like knotholes. This also contributed to the pile of trim loss, as well as shorter rolls.

After finding this out, the team really went to work. They took measurements on how trim loss was generated—bad rolls, bad spots, bad cutting. They found that the foreman contributed little to the trim loss. In fact, when the foreman found a bad spot in a roll, he would adjust his cutting positions at that time and, thus, was able to regroup his orders to take advantage of such anomalies.

Our team wrote up their experiences in model building with a recommendation that the cutting process be looked at in detail, and they outlined an approach. It called for instrumenting the cutting machine with electric eyes to detect bad spots and rolls, and tying the measurements to a computer. Then, utilizing linear-programming techniques the team determined how the blades should be set for an individual roll based on the remaining orders. The blades would be automatically set by the computer. The cost was estimated, and a preliminary cost-effectiveness study was made, comparing the present mode of operation and this proposed technically advanced method. The team concluded with a strong recommendation for keeping the foreman.

EPILOGUE

It has been noted that the human trained in dealing with reasonably sized optimization problems can both solve his problem and come rather close to the optimum. This was the case with Stigler and the diet problem, and our foreman. As the problems become more complex, the human tends to lose out as the power of the mathematical model becomes overwhelming. But not always.

THE PERSONNEL-ASSIGNMENT PROBLEM

Never volunteer.
ANONYMOUS

In the old days of World War II, a draftee was given a variety of aptitude tests to determine the most suitable matching of his skills and the Army's needs. Tests were taken to measure his ability in radio, mechanics, electricity; the scores were analyzed; needs were studied; and he found himself in the infantry.

Today, things are different. Tests are still taken, needs are still analyzed, but there is a good chance that the draftee will end up training for a military career which does require abilities which have some resemblance to his aptitude profile. The problem of matching job requirements to available manpower resources has been a continuing research program of the Army, as well as industrial organizations. Linear programming has been an important research tool in the field of personnel classification. The personnel-assignment problem is central to this research, and it can be stated and solved as a linear-programming model.

Let us look in at a draftee center when business is slow. Only three draftees are being processed by the center at Camp LP. For obvious reasons, we shall name them Able, Baker, and Charlie. They have taken a series of tests to determine their aptitude for careers as a radioman, computer programmer, and clerk. Their scores are shown in the following array:

	Radio	Computer	Clerk
Able	5	4	7
Baker	6	7	3
Charlie	8	11	2

The higher the score, the higher the aptitude of the draftee for the corresponding job. Charlie would probably do really well as a computer programmer, but be a dud as a clerk. The draft center has been given the quota for three men—our Able, Baker, and Charlie—and the center must assign them to three available jobs: one each in radio, computer, and clerical schools. The problem faced by the center is how should the assignments be made in order to maximize the utility of the new draftees' service to the Army.

The linear-programming and psychological-testing approach to the solution of this problem is to assume that the scores on the tests enable us to measure the worth of a man if he is assigned to the corresponding job and that the total worth of all the assignments is the sum of the scores. For example, if Able is assigned the first job, Baker the second, and Charlie the third, as shown by the assignment table

	Radio	Computer	Clerk
Able	1	0	0
Baker	0	1	0
Charlie	0	0	1

then the total score is $5 + 7 + 2 = 14$. We may argue the validity of measuring the total worth in this manner—we are assuming a linear relationship between the worth of the individual assignments. We leave it to the reader who is disturbed by the use of these psychological measures to pursue the matter in the literature.

The above table is called an assignment table as, rightly enough, it assigns one man to only one of the jobs—each man to a job and each job to a man. Thus, an assignment table is a square array with exactly one one in each column and row. Another assignment for the three draftees is given by

	Radio	Computer	Clerk
Able	0	1	0
Baker	1	0	0
Charlie	0	0	1

This assignment has a value of $6 + 4 + 2 = 12$. For our 3×3 problem, there are only $3 \times 2 \times 1 = 6$ possible assignments. It is easy to exhaust all the possibilities and show that the assignment

	Radio	Computer	Clerk
Able	0	0	1
Baker	1	0	0
Charlie	0	1	0

yields the maximum score of $6 + 11 + 7 = 24$. When the number of people to be assigned becomes rather large, such an exhaustive treatment becomes impractical. However, a linear-programming model for this application can be developed in a rather straightforward manner.

Here it would help to consider the problem formulation in the light of our knowledge of transportation problems. In essence, we have a number of origins called Able, Baker, and Charlie; each such origin has one unit of material—a draftee—to be shipped to some destination. The destinations—radio, computer, and clerk—each require one unit of the available 3 units. Unlike the transportation problem, we wish to determine the shipping pattern which *maximizes* the total shipping cost—in reality, the sum of the test scores for the assignments. Arranging the data in a tableau similar to the transportation problem, we have

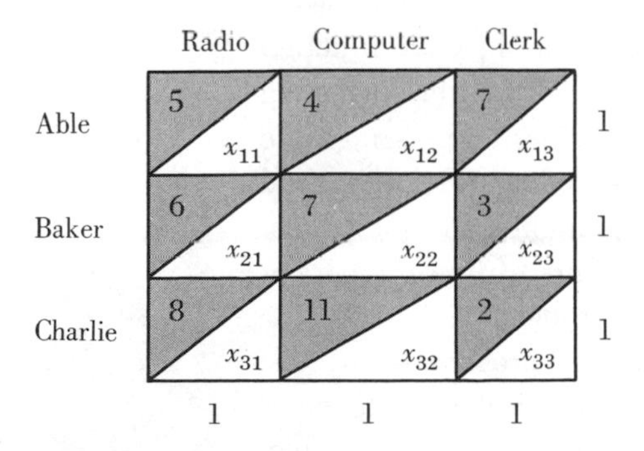

The variables are interpreted as an assignment of the corresponding man to the corresponding job; that is, x_{23} represents the assignment of Baker to the clerical school. Each variable must be positive or zero, and in fact, each variable is restricted to be either a one or a

zero. The optimum solution given above states that $x_{21} = 1$, $x_{32} = 1$, $x_{13} = 1$ and all the other x's are zero. Taking our lead from the transportation problem, the linear-programming model of the personnel-assignment problem is to find nonnegative values of the variables, $x_{ij} \geq 0$, which maximizes

$$5x_{11} + 4x_{12} + 7x_{13} + 6x_{21} + 7x_{22} + 3x_{23} + 8x_{31} + 11x_{32} + 2x_{33}$$

subject to the constraints

$$\begin{array}{lllllllllll}
x_{11} + x_{12} + x_{13} & & & & & & & & = 1 \\
& & & x_{21} + x_{22} + x_{23} & & & & & = 1 \\
& & & & & & x_{31} + x_{32} + x_{33} & & = 1 \\
x_{11} & & + x_{21} & & & + x_{31} & & & = 1 \\
\quad x_{12} & & & + x_{22} & & & + x_{32} & & = 1 \\
\qquad x_{13} & & & & + x_{23} & & & + x_{33} & = 1
\end{array}$$

The solution to this problem has an explicit requirement that variables take on integer values—either zero or one. Fortunately, the mathematical structure of the problem is the same as the transportation próblem, and we are guaranteed an optimal solution in integers. In the trim problem, we made no mention of integer requirements. The trim model can yield an optimum solution which calls for the application of a blade setting a fraction of a time. For manufacturing problems, there is usually no harm done if the fraction is rounded up—it just costs a little more. When assigning a given number of people, however, we must have a solution in integers.

Some assignment problems might require an objective function which is to be minimized instead of maximized. If we were assigning men to different types of work and the "test values" represented the time it took a man to perform the corresponding job, we would want to determine the assignment which minimized the total time. In any event, the same computational procedure of linear programming can solve either type of optimization problem.

An interesting variant of the assignment problem—and one which has been used to demonstrate the sociological importance of linear programming—is the marriage problem. Here, for example, we have 100 men and 100 women, and we wish to pair them off so that the total "happiness" of the pairings is maximized. We need to know a numerical score which measures the happiness of a coupling for each of the possible hundred pairings of a man with a woman. If we denote by x_{ij} the fraction of the time that the ith man spends with the jth woman and if our objective is to maximize overall happiness, then the mathematical model is just a large assignment model. As noted earlier, optimum solutions to such models are in terms of integers—here either 0 or 1. Thus, although each man is given a chance to fractionalize his time between a number of women, society's happiness is better served if he doesn't philander. As was pointed out by a reporter to George B. Dantzig, the formulator of this problem and the founder of linear programming, we could be working with the wrong kind of models.

THE ACTIVITY-ANALYSIS PROBLEM

Have you heard of the wonderful one-hoss shay,
That was built in such a logical way
It ran a hundred years to a day?
OLIVER WENDELL HOLMES

A modern way of looking at the manufacturing operations of an organization is to describe each item produced by the organization in terms of the amount of resources required to manufacture one unit of the item under consideration. The making of a unit—a unit of production activity—represents the utilization of some of the available resources. The problem of the production manager is to determine how many items of each unit he should produce—the levels of activity—which enable him to maximize the company's profits subject to the restrictions imposed by resources available to him. This rather simplistic approach—the viewing of complex operations in terms of basic interrelated activities—is a powerful method for resolving and understanding a wide range of industrial processes. Inherent in this approach is the linear-programming activity-analysis model. For a discussion of this model, let us look into the monthly board of directors meeting of the Simple Furniture Company—our motto: *Proba mers facile emptorem reperit.*[1] President Simon is just starting to describe the new look for his company's manufacturing technology.

"Gentlemen and, of course, Mother Simon, a recent study of our manufacturing organization conducted by Super Management Consultants has brought new insights into how we can improve the

[1] "Good merchandise finds a ready buyer"—Plautus.

operations of our company. In order to convey an understanding of this new approach in the few minutes allotted to this part of the program, I shall take some liberties and greatly simplify the discussion. I have much of what I am going to relate pictured on these flip-charts.

"On the first chart I have shown the four pieces of furniture that we make out of solid mahogany: chairs, tables, desks, bookcases. You will note that besides the main ingredient of mahogany

wood, some of the pieces use leather; we use sliding glass doors for the bookcases; all use glue, plus a few minor items like screws, which we shall ignore. When I mention the word chair, we all picture one of our lovely designs as it appears on the showroom floor. From a manufacturing point of view, we come up with a different picture—like the one I have on the next chart.

"You see, one chair is really equal to 5 board-feet of mahogany, plus 10 man-hours of labor, 3 ounces of glue, and 4 square feet of leather. We just put these resources together so they look like a chair. A table is equivalent to 20 board-feet, 15 man-hours, and 8 ounces of glue. Our pride and joy—the Simon-designed desk—is really 15 board-feet, 25 man-hours, 15 ounces of glue, and 20 square feet of leather, and it is a beauty. The companion bookcase requires 22 board-feet, 20 man-hours, 10 ounces of glue, and 20 square feet of glass.

"Let us take an average production week. When we start the week off, our v-p for production, Sid Simon, gets a listing of the resources available, does some fast figuring, and tells the boys on the floor how many units of each piece of furniture to make. Now, my purpose here is not to throw Sidney out of a job, but to give him a decision-making aid which will enable him to increase profits. On this chart I have listed the total available resources for the average week:

20,000 board-feet of mahogany
4,000 man-hours
2,000 ounces of glue
3,000 square feet of leather
500 square feet of glass

"What Sid has to do is to determine how he puts these resources together to make the most money. To do this he gets from our v-p for finance, Harry Simon, the latest profit figures for each unit, which I have listed here:

$ 45 per chair
$ 80 per table
$110 per desk
$ 55 per bookcase

How many units should we make? How can we get the most profit? Is Sid doing a good job? Let me show you how our new linear-programming model will answer these and other questions.

"I have to get a little mathematical, but I will try and use as little as possible. We've all been out of school a long time. I want to represent the number of units to make of a piece by the first letter of its name, c for chair, t for table, and so on. If $c = 10$, I build ten chairs. For each chair I build, I use these many resources

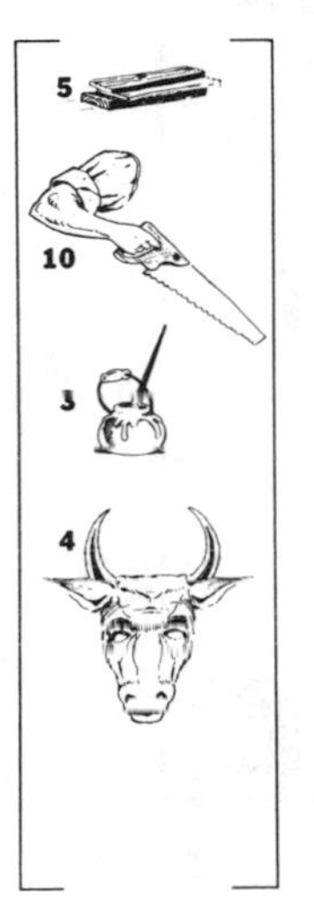

If I build c chairs, I use up c times these resources or

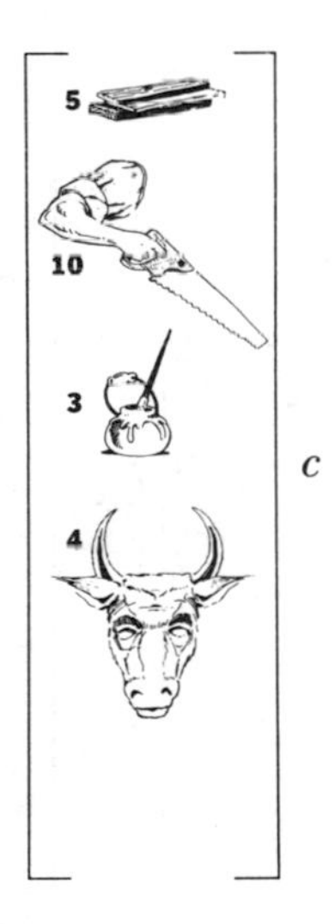

The same thing for the tables, desks, and bookcases. I can let the total amount used up in producing *c* chairs, *t* tables, *d* desks, and *b* bookcases be represented by

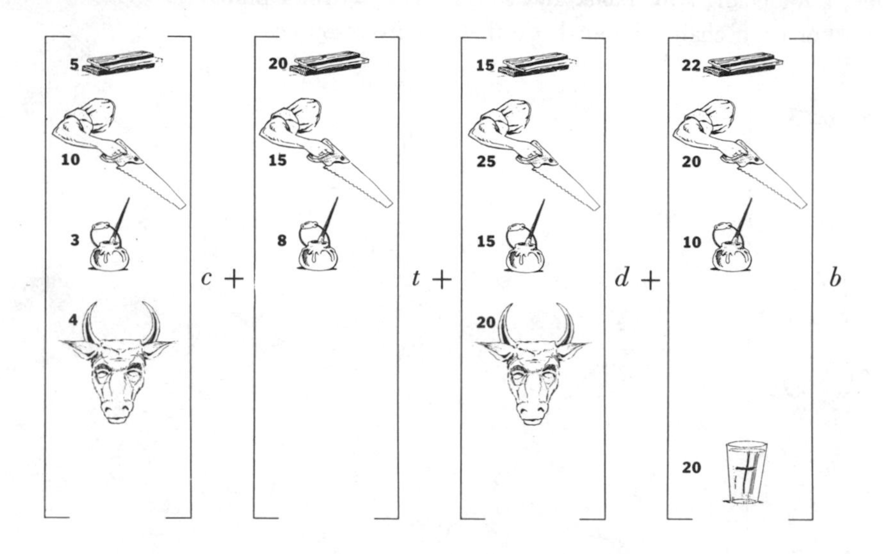

This total amount used must be less than or equal to the total resources available, so I must have

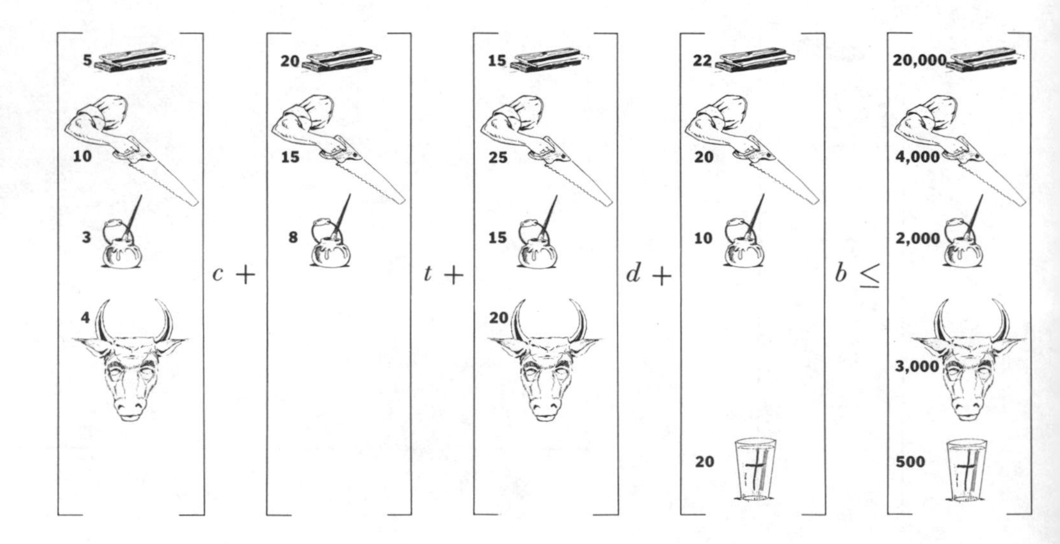

The profit to be made is represented by the sum

$$\$45c + \$80t + \$110d + \$55b$$

All we need to do now is to find the right values of c, t, d, b which makes that sum as large as possible. This is what Sid tries to do. Linear programming guarantees giving us the maximum profit. Any questions? Sid."

"You sure left out a lot of things. I have to make sure we meet the sales forecast. The boys in the field call in with their orders and they want their orders right away. The way you glibly stated the problem I have no guarantee that I will make any chairs, for example. Also, I might end up making too many chairs, or only chairs. How can you take care of these things?"

"Simple Sid. Let me write the inequalities without the pictures like this:

$$\begin{aligned}
5c + 20t + 15d + 22b &\leq 20{,}000 \\
10c + 15t + 25d + 20b &\leq 4{,}000 \\
3c + 8t + 15d + 10b &\leq 2{,}000 \\
4c + 20d &\leq 3{,}000 \\
20b &\leq 500
\end{aligned}$$

If you need at least 50 chairs all you add to the model is the inequality

$$c \geq 50$$

If you don't want more than 30 desks all you add is

$$d \leq 30$$

It's pretty easy. The linear-programming model can take care of all these things, and more. We can bring in the transportation, inventory and warehousing problems and attempt to optimize the whole Simple System. That's what we are going to work on next. No more questions—time's up."

SOLUTION OF PROBLEMS

{We combine some age-old concepts of Euclid with twentieth-century intuition to understand the process behind the solution of a linear-programming problem.}

THOSE READERS who are at all concerned about solving a particular linear-programming problem can lay their cares aside and move on to the next chapter. We can solve just about any linear-programming problem. There are computational procedures which enable us to solve small-scale problems by hand. Even for the big (and not so big) problems, the same procedures, when coupled with an electronic computer, have been forged into a major weapon of the problem-solver. The first thing most computers "learn" after they have been imbued with the basic rules of arithmetic is how to solve linear-programming problems. This phenomenon attests to the pervasiveness and power of the linear-programming model.[1]

I hope that few readers will turn away from the main body of this chapter. Although one can understand a great deal about linear programming without knowing much about solving a problem—one can even make a living applying such limited knowledge—it is axiomatic that a basic and insightful comprehension of this field can be gained only by a careful blending of the theoretical, computational, and applied aspects. As I am purposefully ignoring part of this prescription—the theoretical side of the house—it is felt that the "house" and main objective of this book would not hold up to inspection without some discussion of the computational considerations. Thus, I shall present a limited discussion of the solution process associated with linear programs so that the reader can develop a deeper insight and appreciation of the total subject.

To facilitate the inquiry, we must standardize the required mathematical notation. We shall, in general, use very simple numerical examples whose mathematical models require two variables. These variables are denoted by the shorthand symbols x_1 and x_2. From a plane (Euclidean) geometry point of view, x_1 represents a dimen-

[1] As the initial developments in the fields of electronic computers and linear programming evolved at about the same time period—late 1940s—and since these developments have been mutually beneficial, it is unfortunate that a serious confusion in terminology occurred. We talk about computer programming, which deals with the logical analysis and the set of machine instructions to solve a given problem. Thus, we have a computer program to solve a linear-programming problem. As a slight historical note, the U.S. Air Force Project SCOOP (Scientific Computation of Optimal Programs), under the direction of G. B. Dantzig and M. K. Wood, gave early support to the development of electronic computers. The Air Force aided in the development of the National Bureau of Standards SEAC Computer (1950) and the problem solved at the dedication of the SEAC was a linear program. This Air Force group also obtained the second UNIVAC computer for the specific use of solving Air Force programming problems (1952).

sion, or direction, and x_2 represents a second dimension, or direction. Mathematically, we say we are working in two-dimensional Euclidean space. Geometrically, we can illustrate our space and dimensions with the familiar two-axis graph—but more of this later.

In the examples of Chapter 2, the variables took on different meanings, depending on their definition for the problem under consideration. In the trim problem, the x's represented settings of the cutting blades. For some of the problems we identified the variables after certain names, like K for Krunchies. We shall rename all variables with an x and an appropriate numerical or letter subscript, The generic variable is given the identification symbol x_j. As some of the problems we shall solve are just numerical and symbolic "toy" examples, we should get used to dealing with variables which do not have a specific meaning. Thus, we can talk about the equation $x_1 + x_2 = 1$ without having to give these variables a particular interpretation.

Although most of the examples discussed in this chapter involve only two variables, the reader should not have any difficulty extending the concepts and notions developed for these examples to situations which involve a large number of variables. The trim problem required seven variables or dimensions, and its formulation required no Einsteinian expansion of our dimensional experiences and intuition. The space into which we are about to tread is accessible to all.

SINGLE-VARIABLE PROBLEMS (FOR THE TRUE BEGINNER)

The distance is nothing; it is only the first step which counts.
MADAME DU DEFFAND

The simplest of all optimization problems is to find the maximum value of the single variable x_1, where x_1 is not subject to any restrictions. Here, the numerical value of x_1 can become as large as we want, and we say that the maximum value is unbounded. Geometrically, we are looking for the largest value of x_1 along the infinite number line or axis. As we have no restrictions on x_1, we can march

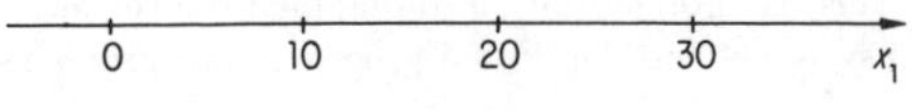

0 10 20 30 x_1

along this line as far to the right as we care to without encountering any barriers. By restricting x_1, the problem becomes more interesting, but still easy to solve. We shall deal with variables whose values are always restricted by the basic linear-programming requirement of nonnegativity—thus, $x_1 \geq 0$.

The problem of maximizing x_1 subject to the inequality constraint $x_1 \leq 20$ allows us to move along the x_1 axis from the point $x_1 = 0$ until we encounter the barrier—a stop sign—at the point $x_1 = 20$.

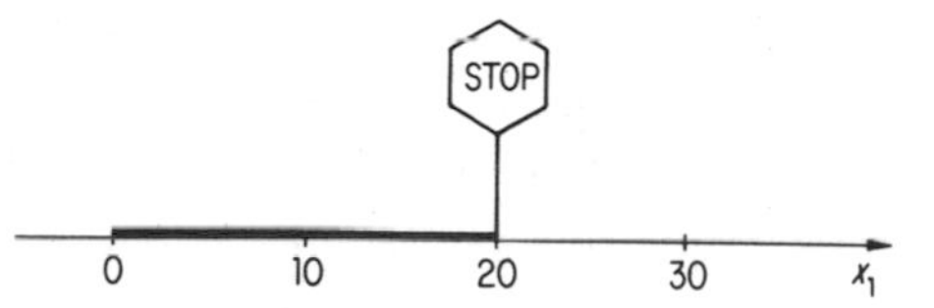

Any point on the line connecting $x_1 = 0$ and $x_1 = 20$ represents a potential solution to the problem. This set of points is called the solution set, or solution space. As we are maximizing, the optimum point in the solution space of the problem—the point which makes the objective function, maximize x_1, as large as possible—is $x_1 = 20$. If we were looking for the minimum value of x_1 subject to $x_1 \leq 20$, the optimum answer—remembering the nonnegativity restriction—would be $x_1 = 0$.

Changing the inequality restriction of the maximizing problem to $2x_1 \leq 20$ restricts our freedom of movement along the x_1 axis to the points between $x_1 = 0$ and $x_1 = 10$, with the optimum answer

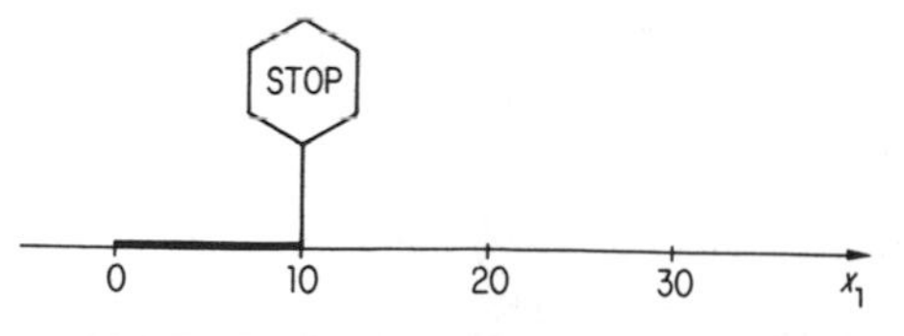

being $x_1 = 10$. Single-variable, single-restriction problems can readily be solved by such a simple analysis. If the single restriction is an equation like $x_1 = 20$ or $2x_1 = 20$, then the solution space is just a single point—here, $x_1 = 20$ and $x_1 = 10$, respectively. Linear-programming problems can involve equality as well as inequality restrictions, but the geometric basis of such problems is made sharper if we stress the inequality nature in our formulations.

A typical example of a single-variable linear-programming problem with many restrictions is to maximize x_1 subject to

$$5x_1 \leq 75$$
$$6x_1 \leq 30$$
$$x_1 \leq 10$$

with, of course, $x_1 \geq 0$. We need to find the largest value of x_1 which satisfies the three inequalities simultaneously. We first determine the solution space associated with the individual inequalities and then find the solution space which represents the joint solution of all the inequalities.

For $5x_1 \leq 75$ we have that any value of x_1 between 0 and 15 satisfies the constraint.

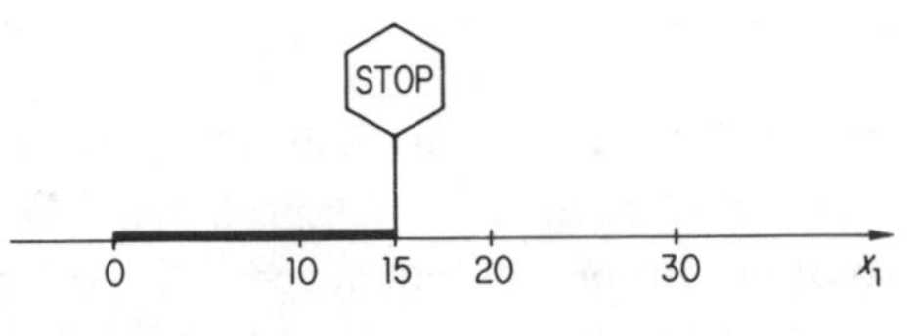

For $6x_1 \leq 30$, the solution space is between 0 and 5.

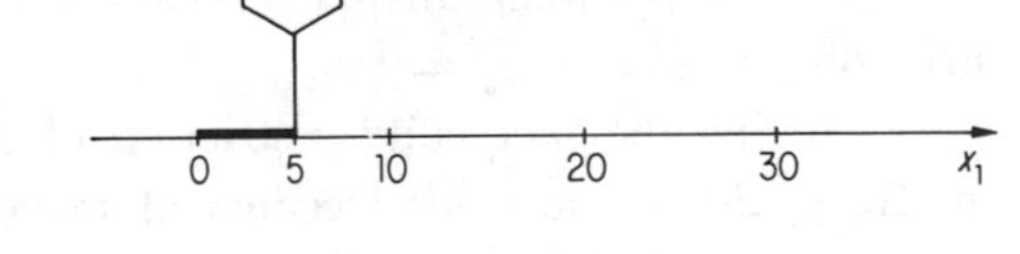

And finally, for $x_1 \leq 10$ the solution set is from 0 to 10.

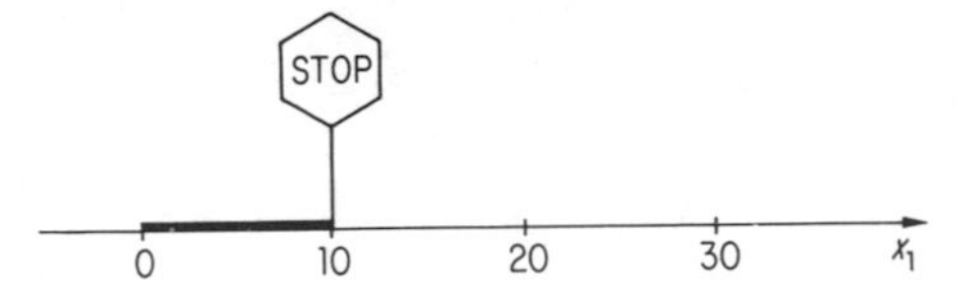

If we combine all the stop signs onto one graph we have

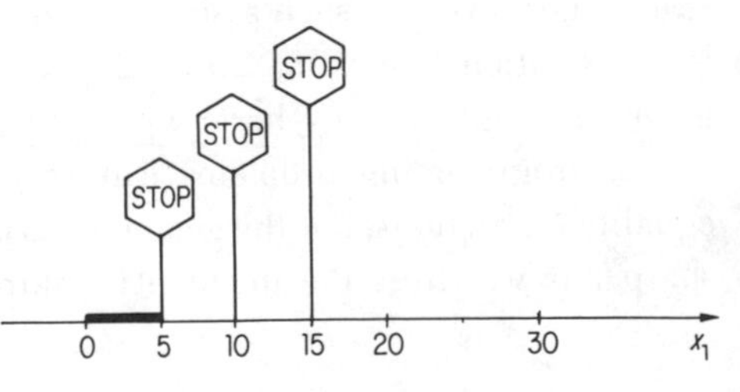

Thus, our march from the point $x_1 = 0$ to the right cannot get by the point $x_1 = 5$. Any value between 0 and 5 will satisfy the three inequalities simultaneously—this set of points is the solution set of the problem. As we are looking for the largest point in the solution set, the optimum answer to the original three inequality problem is $x_1 = 5$. For any problem in one variable we can determine the optimum solution as illustrated above.

TWO-VARIABLE PROBLEMS

Every step forward in the world was formerly made at the cost of mental and physical torture.

NIETZSCHE

We next enter the two-dimensional problem arena, and it is here that the challenge to solve linear programs becomes nontrivial and more instructive. To study this area I must make certain assumptions concerning the reader's mathematical sophistication. What follows requires knowledge about graphing two-dimensional constraints—inequalities and equalities—and demands of the reader some understanding of basic high school algebra. Although I shall assume this level of knowledge, the presentation will be as rudimentary as possible so that all readers can enter the arena with sufficient arms.

The two variables or dimensions x_1 and x_2 are represented by the right-angled axes

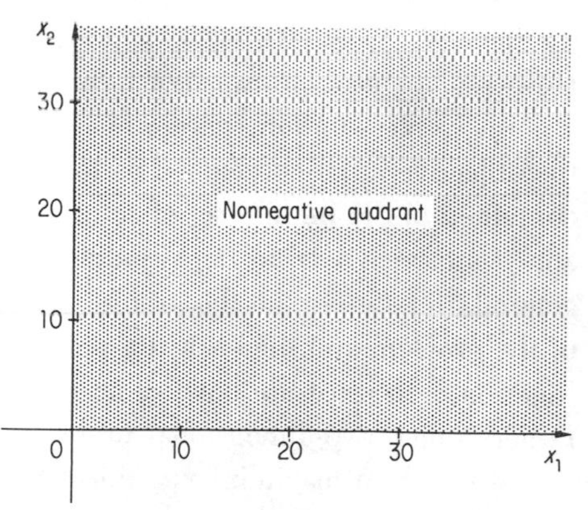

Since we are restricting the range of the variables to the nonnegative set of values, the associated graph is constrained to certain directions. For x_1, we can only move to the right of the point 0, the

origin of the axes, and for x_2 we can only move straight up from the origin in a direction perpendicular to the x_1 axis. Mathematically, we say that the solution space is restricted to the nonnegative quadrant.

A value of $x_1 = 20$ sends us 20 units to the right, and a value of $x_2 = 30$ marches us 30 units straight up. Taken together, these directions put us inside the nonnegative quadrant as shown.

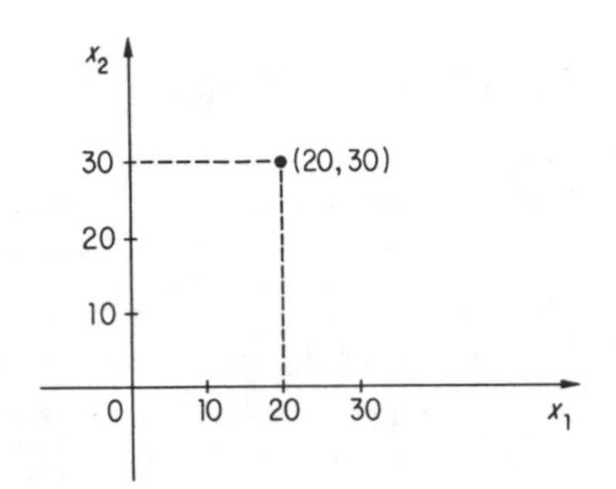

We use the shorthand notation $(x_1, x_2) = (20, 30)$ to represent a point on the two-dimensional graph.

In one dimension the inequality $x_1 \leq 20$ has a solution space which ranges from $x_1 = 0$ to $x_1 = 20$. Any point in this range satisfies the requirement of the inequality—the nonnegative variable x_1 must be less than or equal to 20. Thus, there are an infinite number of solutions contained in this solution space. To list a few, we have 0, 1, 2, 10, $\sqrt{3}$, and $3/4$, as permissible values of x_1. However, the problem was not to find the solution space, but to find a point in the solution space which maximizes x_1. The addition of the objective function enabled us to "home in" on a particular point contained in the infinite set of solution points. This same graphical analysis carries over to the two-dimensional problem. We must, in general, first determine the infinite set of solutions to the constraints of the problem and then select a point which optimizes the objective function. To illustrate how we accomplish this, we next solve some rather simple two-dimensional problems.

We wish to maximize x_1 subject to the conditions $x_1 \leq 20$ and $x_2 \leq 30$, with $x_1 \geq 0$ and $x_2 \geq 0$. In the two-dimensional graph we plot the barrier lines $x_1 = 20$ and $x_2 = 30$.

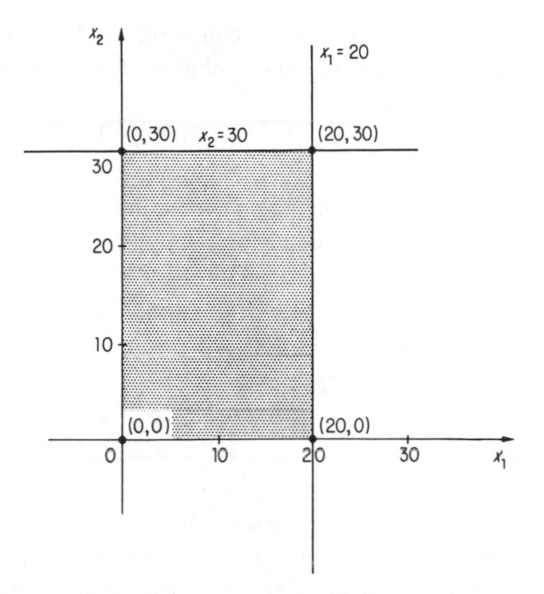

The nonnegativity conditions $x_1 \geq 0$ and $x_2 \geq 0$, taken together with the barrier lines, force us to confine our search for solution points to the shaded area and its boundaries—this is the solution space. No matter where we look in the solution space the largest value of x_1 is 20. But here we notice a strange thing—there are many points in the solution space which yield a value of $x_1 = 20$. The points (20, 0) and (20, 30) are both in the solution space and have $x_1 = 20$. The optimum solution is not unique—a rather common occurrence in linear programming, one which offers us no additional concern. In fact, we really have an infinite number of optimum solutions, as any point on the line segment joining the points (20, 0) and (20, 30) is a solution point with $x_1 = 20$.

An important peculiarity of linear-programming problems is that well-behaved problems, like the one under discussion, have optimum-solution points which include the vertices formed by the intersection of the boundary lines. The present problem has four such vertices—(0, 0), (0, 30), (20, 0), (20, 30)—with the latter two being optimum solutions for the objective function maximize x_1.

If we change the objective function to maximize x_2, we have the optimum vertices (0, 30) and (20, 30), along with any point on the line segment joining these vertices. If we wanted to minimize x_1 or minimize x_2, we find the former objective function minimized by the vertices (0, 0) and (0, 30); while the latter is minimized by (0, 0) and (20, 0). Of course, points on the lines connecting the pairs of vertices are also optimum solutions for the respective objective functions.

Let us complicate the problem by changing the objective to finding the values of x_1 and x_2 such that the sum $x_1 + x_2$ is minimized, subject to the same constraints. This objective function takes on its smallest value at the point (0, 0)—that is, minimum $x_1 + x_2 = 0$. This optimum solution is unique in that any other point in the solution space has a positive value for at least one of the variables.

What about the solution if the objective function was to maximize $x_1 + x_2$? A slight search into the possibilities reveals the one solution (20, 30) for a maximum of $x_1 + x_2 = 50$.

By this time the reader might have rightly concluded that no matter what the objective function, the optimum solution will occur at one of the four vertices and possibly on some other boundary points. This is true for all well-behaved linear-programming problems—an ill-behaved problem being one in which the objective function is unbounded, as was the case for our first, single-variable problem. At this point it would be correct to inquire what all the fuss is about. If the optimum solution to a linear-programming problem occurs at one of the vertices, all we need do is to determine all the vertices, evaluate the objective function for each vertex, and pick out the one that optimizes the objective function. For example, let the objective function be to maximize $3x_1 - 2x_2$ and set up the table

Vertex	Value of $3x_1 - 2x_2$	Maximum
(0, 0)	$3x_1 - 2x_2 = 3(0) - 2(0) = 0$	
(20, 0)	$3x_1 - 2x_2 = 3(20) - 2(0) = 60$	60
(0, 30)	$3x_1 - 2x_2 = 3(0) - 2(30) = -60$	
(20, 30)	$3x_1 - 2x_2 = 3(20) - 2(30) = 0$	

The optimum value is 60, and the unique optimum solution point is (20, 0). The reader can do the same for any linear objective function involving x_1 and x_2, subject to the given constraints. This looks rather easy, and it is for such toy examples. The difficulty in this approach for the general problem is that the number of vertices can be immense, and their enumeration is quite a computational task. The above approach is acceptable for the discussions in this chapter. Even here, however, we shall refine the process to bring it more in line with the actual algebraic computational process used to solve linear-programming problems, the *simplex method.* To do this, we must attack more involved problems. But first, we shall introduce some of the basic terminology of linear programming.

The new math-educated readers must have recognized that the solution space of the two-variable problem is a *convex set* of points and that the vertices are *extreme-points* of the convex set. A set of points is convex if the set contains the entire line segment joining any two of its points. The following are convex sets

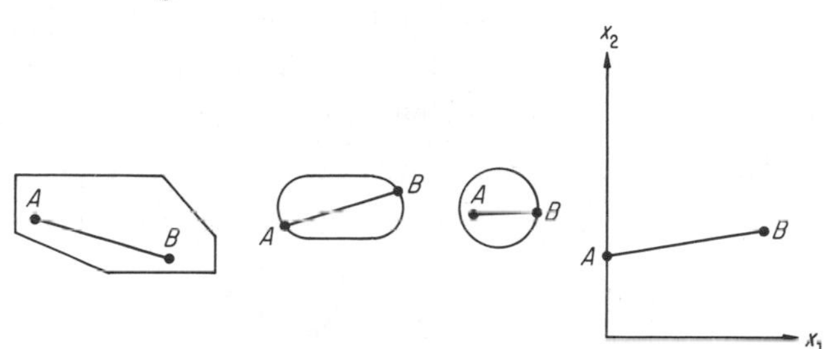

while the following are not convex sets

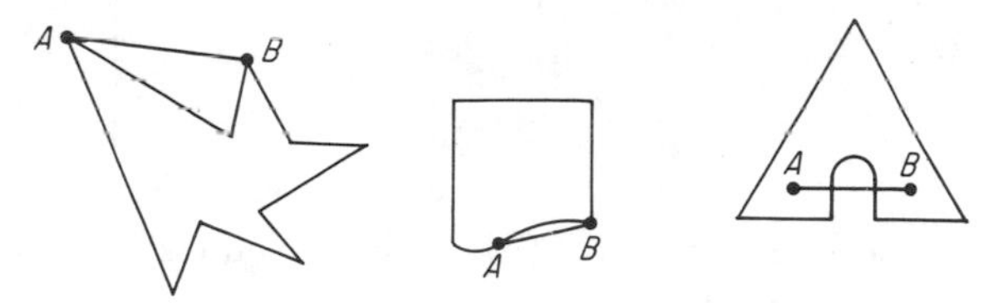

A point of a convex set is an extreme point if it does not lie on a line segment joining two other points of the set. A convex set can have a finite number, infinite number, or no extreme points. We leave it to the reader to supply examples of each type. The convex set of a linear program has a finite number of extreme points. In general, such a convex set is a *convex polyhedron.*

A MANUFACTURING PROBLEM

It is a bad plan that admits no modification.
PUBLILIUS SYRUS

We next consider a smaller version of the manufacturing problem encountered by the Simple Furniture Company. To reduce it to a two-variable problem, we shall consider only the making of chairs and tables, subject to the board-foot and man-hour restrictions, and change the amount of the available resources to facilitate the discussion of a graphical solution. Our problem then is to maximize the profit function $\$45x_1 + \$80x_2$ subject to

$$5x_1 + 20x_2 \leq 400$$
$$10x_1 + 15x_2 \leq 450$$

Here, x_1 stands for the number of chairs (c) to be manufactured and x_2 for the number of tables (t), with $x_1 \geq 0$ and $x_2 \geq 0$. We have a total of 400 board-feet of mahogany and 450 man-hours to combine into a manufacturing schedule for chairs and tables.

The first thing to be done is to delineate the solution space. We start out with the solution space restricted to the nonnegative quadrant

The first inequality, $5x_1 + 20x_2 \leq 400$, requires us to stay in that part of the nonnegative quadrant whose points (x_1, x_2), when combined as required by the inequality, yields a sum less than or equal to 400. The easiest way to determine this region is to plot the barrier line $5x_1 + 20x_2 = 400$. We do this by finding the two points—one on the x_1 axis and one on the x_2 axis—through which the line passes. If $x_1 = 0$, we must have $x_2 = 20$; if $x_2 = 0$, we have $x_1 = 80$. The two points (80, 0) and (0, 20) lie on the barrier line

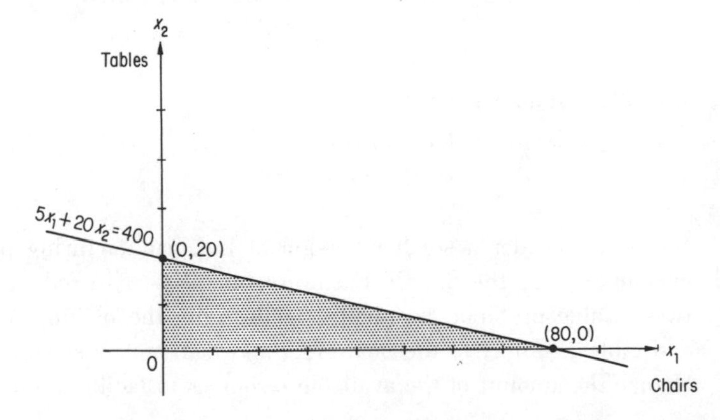

All points in the shaded area and its boundary satisfy the inequality $5x_1 + 20x_2 \leq 400$.

For the second inequality, the barrier line is $10x_1 + 15x_2 = 450$, and the intersection points on the axes are (45, 0) and (0, 30). This is shown by

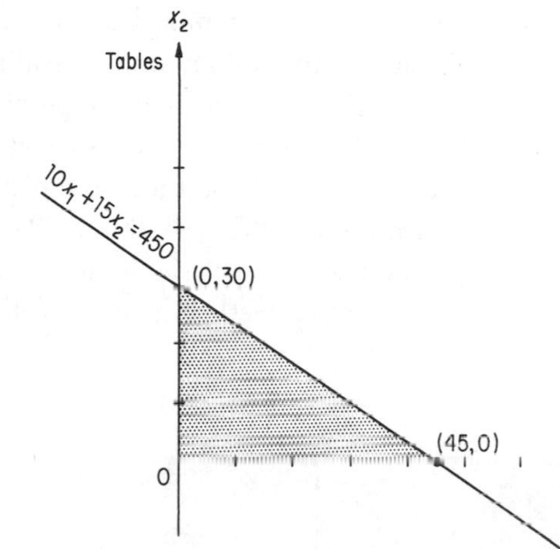

Any point in this convex set satisfies $10x_1 + 15x_2 \leq 450$. The joint solution of the two constraints is found by superimposing both shaded areas onto one graph—the areas which overlap form the required solution space.

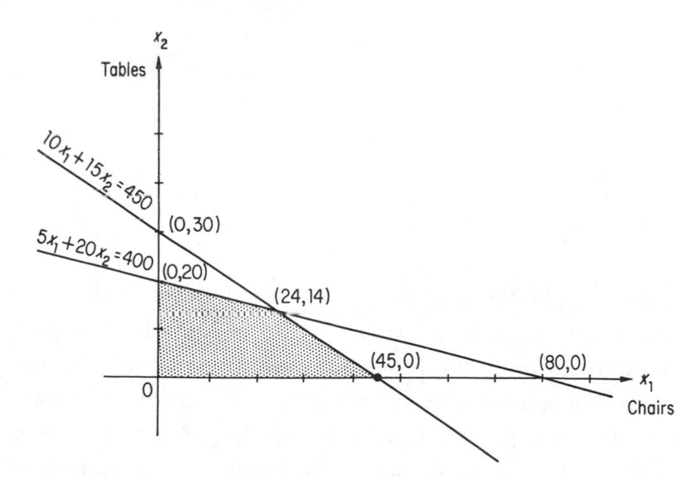

This combined solution space represents the many combinations of chairs and tables which can be manufactured subject to the availability of 400 board-feet and 450 man-hours. Only points in this region satisfy the constraints of the problem.

Up to now we have identified all the extreme points of the convex set of solutions except the one generated by the intersection of the two barrier lines. This point represents the manufacturing of an amount of chairs and tables which uses up all the available resources. Any point (x_1, x_2) on the line $5x_1 + 20x_2 = 400$, e.g., (0, 20), uses up exactly 400 board-feet. Similarly, any point on the line $10x_1 + 15x_2 = 450$ requires the full amount of available man-hours. By a little algebraic juggling we find that the point $x_1 = 24$ chairs and $x_2 = 14$ tables satisfies both constraints exactly.

The problem is to determine which of the infinite number of points in the shaded region maximizes $45x_1 + 80x_2$. We can treat this expression like an equation with the variables x_1 and x_2. For example, the equation $45x_1 + 80x_2 = 0$ is shown by the dashed line

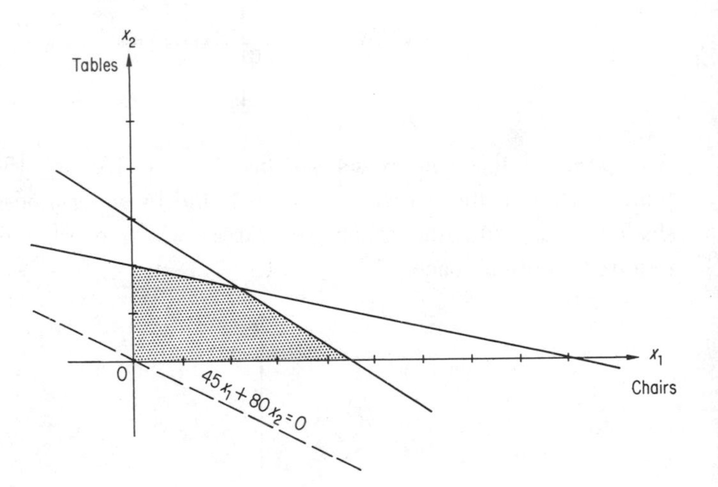

If the Simple Manufacturing Company does not make any chairs or tables, i.e., $x_1 = 0$, $x_2 = 0$, this profit line shows a gain of zero. We want to move the profit line in a direction which increases the profit, but subject to the restrictions of the available resources. We can slide the profit line into the shaded area and keep moving it until we are stopped by one of the barriers—we cannot look outside the shaded area for any solutions. Such a movement of the profit line is illustrated by

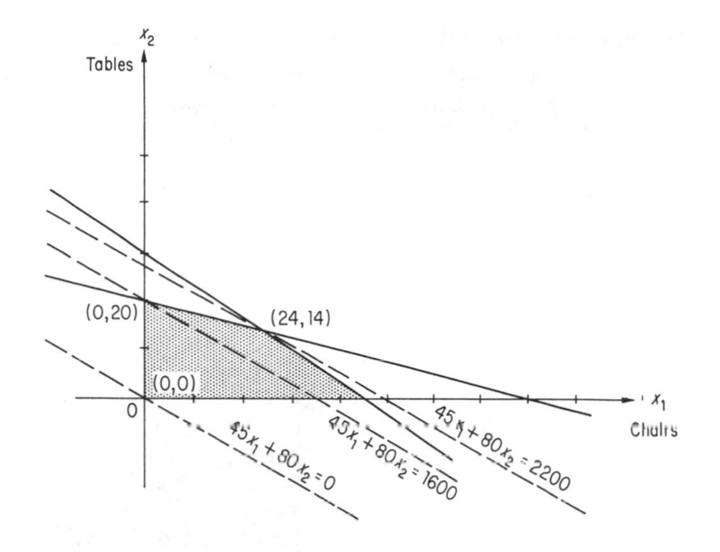

The optimum manufacturing plan is to make $x_1 = 24$ chairs and $x_2 = 14$ tables, which yields a profit of $\$45 \times 24 + \$80 \times 14 =$ \$2,200. The sweeping of the objective function through the convex set of solutions is, in effect, what happens when we solve the problem using the algebraic techniques of the simplex method.

The simplex process starts at any extreme point like $x_1 = 0$, $x_2 = 0$, determines that there are more profitable extreme-point solutions, and tries one out for size. The process keeps improving the solution by shifting from one extreme point to a better one—here to $x_1 = 0$, $x_2 = 20$—and continues in this manner until it can demonstrate that the current extreme point cannot be beat. The saving feature of the simplex method is that it is able to pick and choose its path from one extreme point to a better, adjacent, extreme point by trying out a small subset of all the possible extreme points. In this respect, the process is quite efficient, but no one knows exactly why.

The point (24, 14) represents an optimum production schedule for a wide range of objective functions. As the profit for a chair or a table changes, the slope of the profit line also changes—it can be visualized as pivoting about the point (24, 14). A big enough change in the profit coefficients will twist the profit line so that the optimum is at a different point. For any profit line, the optimum would occur at one of the four extreme points. Multiple optimum solutions arise when the objective function takes up a final position which includes a boundary of the convex set. The objective function $\$50x_1 + \$75x_2$ is optimized by the extreme points (24, 14) and (45, 0) and all the points on the boundary joining them.

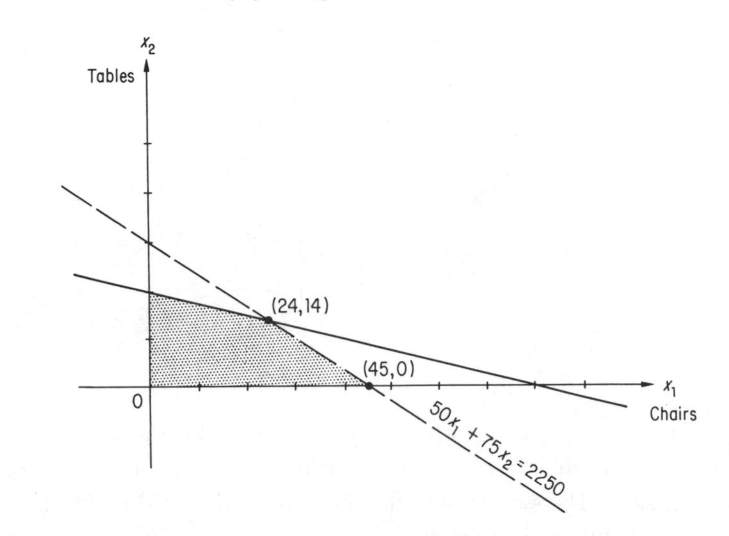

The above approach can be used to solve any two-variable inequality problem. The only points which we are required to calculate are those needed to graph the barrier lines and the extreme point at which the objective function makes its last point of contact with the convex set of solutions.

The same general graphical interpretation applies to problems with higher dimensions. We have a convex polyhedron defined by the constraints, and we wish to pass the objective function through the polyhedron in the direction which yields the optimum. The reader can picture this process by viewing the room he is sitting in as a convex set defined by the constraints of a linear program. One corner of the room is the origin of the three-dimensional world. The floor, walls, and ceiling are barrier constraints, with the convex set being a part of the nonnegative orthant of three-dimensional space. Any objective function, such as maximize $x_1 + x_2 + x_3$, is optimized at one of the eight extreme points. If we remove the ceiling constraint, then the convex set is unbounded. If x_3 is the vertical direction, then the objective function maximize x_3 is unbounded. We leave it to the reader's imagination to conjure up the wide variety of situations that can arise in the three-dimensional world. Again, we emphasize that problems with more than two variables are not solved graphically, but many of them can be solved with pencil and paper and the simplex method.

The reader has probably noted that the optimum solution to the chair-table-manufacturing problem, make 24 chairs and 14 tables, used all the available resources. For this solution the mahogany inequality constraint $5x_1 + 20x_2 \leq 400$ is an exact equality, $5(24) + 20(14) = 400$. Similarly, for man-hours we have $10 \times 24 + 15 \times 14 = 450$. A natural question to ask is what would happen to the company's profits if it could increase the available supply of a particular resource? Also, for the given resources, if the Simple Company now considers the making of desks, as well as chairs and tables, how can it decide whether or not it is profitable to switch its resources to desk production? These types of questions and their answers are basic to economic theory and the running of an industrial organization. The former concerns what is termed marginal analysis, while the latter deals with opportunity costs. For linear-programming models like the chair-table problem—in fact, for any linear-programming problem—these questions can be answered by solving a related linear-programming problem called the dual problem; the original problem is the primal. The dual problem is formed using the data from the primal. For the chair-table problem it is as follows: minimize the resource valuation function $400w_1 + 450w_2$ subject to

$$\begin{aligned} 5w_1 + 10w_2 &\geq 45 \\ 20w_1 + 15w_2 &\geq 80 \\ w_1 \qquad\quad &\geq 0 \\ w_2 &\geq 0 \end{aligned}$$

Here w_1 is the unknown value (accounting price) of the mahogany resource, and w_2 is the unknown value of the manpower resource. (The reader should note the switch between the objective-function coefficients and the right-hand sides of these problems. He should also note that the rows of the primal coefficients are the column coefficients of the dual.) As both these resources are bottlenecks in the optimal solution—i.e., they are scarce and limit the total production—we would expect these dual values to be positive in the optimal solution to the dual. The optimal solution is $w_1 = 1$ and $w_2 = 4$, with a value of the objective function of \$2,200, assuming that the physical units of the w's are measured in dollars per unit of resource. We interpret this answer as follows.

The marginal-profit contribution of an additional unit of mahogany is 1; for manpower it is 4. If one of the resources was not scarce, i.e., an excess of mahogany, then an increase in its availability would not cause a change in profits, and its marginal value would be zero. The main theorem of linear programming (the duality theorem) states that if the primal has a finite optimum solution, so does the dual, and the optimum values of the objective functions are equal. Thus, we note that the minimum valuation of the resources is equal to the total profit of $2,200.

For the primal problem, since the optimal solution requires us to produce both chairs and tables, the opportunity costs are zero. This is reflected in the optimal solution to the dual by the fact that $w_1 = 1$ and $w_2 = 4$ make both the dual constraints an equality. If, for example the optimum solution required the production of only tables, then the opportunity cost for chairs, given by the difference between the left- and right-hand sides of the first inequality, $5w_1 + 10w_2 - 45$, would be positive; i.e., the value of the resources that make a chair are worth more than the profit from a chair. Thus, resources required to make a chair are more valuable when applied to the making of tables. A similar analysis can be made to determine if it would be more profitable to make a third product like desks. The reader should be able to solve the dual of the chair-table problem by graphical techniques and answer the question: If the profit of a desk is $110 and a desk uses 15 units of mahogany and 25 units of manpower, should the Simple Furniture Company make desks?

One of the important contributions of linear programming is its ability to answer such questions. In fact, it is often more important to know opportunity costs and marginal values than the true optimum solution.

THE DIET PROBLEM (AGAIN)

Be plain in your dress, and sober in your diet;
In short, my deary, kiss me and be quiet.

LADY MARY WORTLEY MONTAGU

To illustrate a slightly different geometric situation, we return to the housewife's breakfast problem. We let x_1 stand for Krunchies and

x_2 for Crispies. The mathematical formulation of the problem is to find nonnegative values of x_1 and x_2 which minimizes the total cost

$$3.8x_1 + 4.2x_2$$

subject to

$$\begin{aligned} 0.10x_1 + 0.25x_2 &\geq 1 \\ 1.00x_1 + 0.25x_2 &\geq 5 \\ 110.00x_1 + 120.00x_2 &\geq 400 \end{aligned}$$

Plotting the barrier lines as before, we have the shaded solution space shown below. This convex set is not a convex polyhedron and is unbounded in the direction of increase for both x_1 and x_2.

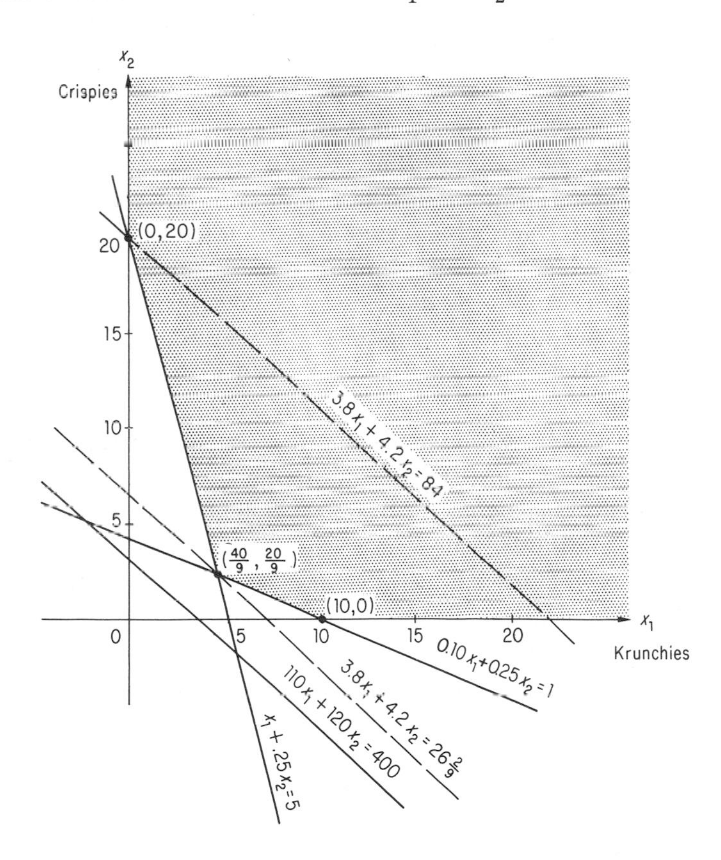

Our graphical analysis has revealed that the constraint on calories will always be satisfied if the other two constraints are satisfied. The barrier line for calories, $110x_1 + 120x_2 = 400$, does not form part of the boundary of the convex set of solutions. It is redundant to the problem and can really be dropped from further consideration. In the figure, we have drawn two traces of the cost line, $3.8x_1 + 4.2x_2 = 84$ cents and the minimum cost line of $3.8x_1 + 4.2x_2 = 26\frac{2}{9}$ cents. The optimum solution is unique and requires the housewife to purchase $x_1 = \frac{40}{9}$ ounces of Krunchies and $x_2 = \frac{20}{9}$ ounces of Crispies.

The dual to the diet problem is as follows: find nonnegative values of w_1, w_2, and w_3 which maximize

$$w_1 + 5w_2 + 400w_3$$

subject to

$$\begin{aligned} 0.10w_1 + 1.00w_2 + 110.00w_3 &\leq 3.8 \\ 0.25w_1 + 0.25w_2 + 120.00w_3 &\leq 4.2 \end{aligned}$$

Here a w_i can be interpreted as the value of one unit of nutrient i. Thus, the dual problem determines the w_i such that the value of the minimal diet is maximal, subject to the constraints that the value of the nutrients in one unit of each food is less than or equal to the cost of the food.

The geometry of linear programming enables us to view the infinite set of solutions as points in the appropriate dimensional space. The solution set is convex, and except for the unbounded optimum, this enables us to confine our search to the finite, but usually large, set of extreme points. Computational procedures like the simplex method restrict the search to a reasonable number of these extreme points. The fascination and power of linear programming is due to the synergistic effect of these efficient computational procedures and the wide range of important applications—one is nothing without the other.

LINEAR POTPOURRI

{A mixture of things linear and otherwise gathered from the fields about Linear Programsville.}

Any attempt to offer a complete appraisal of the topics, problems, and applications which have been influenced by linear programming can only lead to failure. The scope and range of these areas are too wide to be encompassed in a single setting. I have, in the previous chapters, discussed many of the more important, albeit standard, problems from the realm of linear programming. In this chapter, we shall extend these discussions to include additional applications, as well as some of the unusual topics related to linear programming.

NETWORK PROBLEMS

Awake my St. John! leave all meaner things
To low ambition and the pride of kings.
Let us, since life can little more supply
Than just to look about us and to die,
Expatiate free o'er all this scene of man;
A mighty maze! but not without a plan.

ALEXANDER POPE

A rather important class of linear-programming problems falls under the general heading of network problems. As the approach to the formulation of the related mathematical models is rather direct, we shall illustrate it with some small examples.

We are given a transportation network (pipeline system, railroad system, communication links) through which we wish to send a homogeneous unit (oil, cars, message units) from a particular point of the network called the source node to a designated destination called the sink node. In addition to the source and sink nodes, the network consists of a set of intermediate nodes which are connected to each other or the source and sink nodes by arcs of the network. These intermediate nodes can be interpreted as switching or transshipment points. We shall label the source node by 0 and the sink node by m and refer to the intermediate nodes by a number or a letter. A typical small-scale network, which we shall use in the following discussion, is

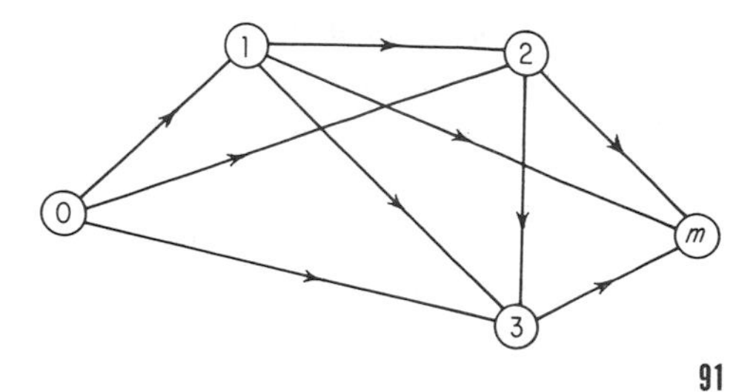

The symbol for an arc connecting nodes 2 and 3 is (2, 3), or, in general, (i, j). The arcs pictured in our network are directed arcs in that the flow of goods along an arc is in the direction of the arrow. If goods can flow both ways, then the corresponding nodes are connected by two arcs with the flows going in the opposite directions.

Each arc of the network can accommodate only a finite amount of flow. A section of a pipeline can pass a designated number of barrels per hour; a communication link can handle so many calls per day, etc. Thus, each arc has a specified upper bound on the flow through the arc which we designate, for example, by f_{23} for the maximum flow through arc (2, 3). In the next figure we show the upper bounds on each arc, here $f_{23} = 1$,

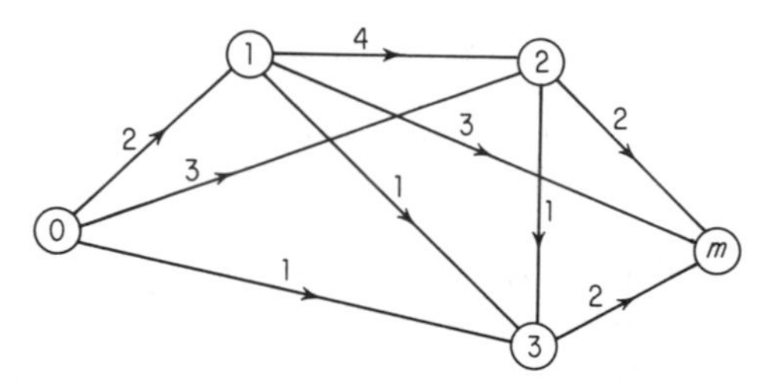

The flow of the commodity which originates at the source 0 is sent along the arcs to the intermediate nodes, then transshipped along additional arcs to other intermediate nodes, or to the sink, until all the commodity which began the trip at node 0 finally arrives at node m. That is, we impose upon the network the condition of *conservation of flow* at the intermediate nodes—what is directed into a node must be directed out. This assumption is analagous to Kirchoff's node law in electrical networks.

Maximal Flow Network Problem

The first problem to be considered—the maximal flow network problem—is concerned with sending as much as possible of the commodity through the network from node 0 to node m. We assume that there is an unlimited supply of the commodity at node 0 and that the only thing which restricts our sending an unlimited supply to node m is that the capacities of the arcs are limited by the given upper bounds. Thus, we wish to determine the maximum amount of flow, designated f, which can be sent from the source to the sink, along with the arcs used, as well as the amount used of each arc's capacity. We wish to direct f through the network as shown

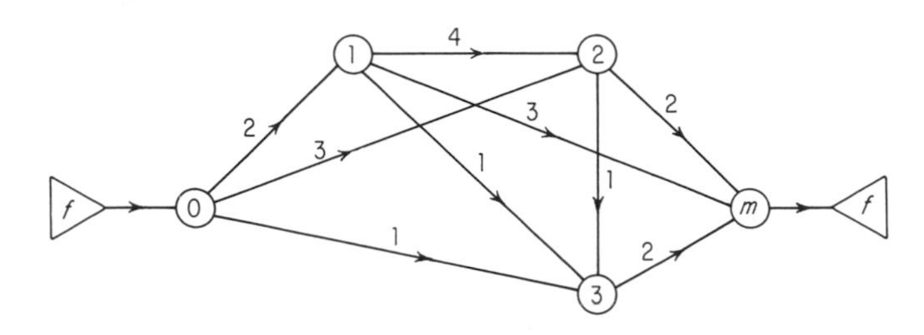

The mathematical model for such a network is easy to come by. By defining

- A variable x_{ij} as the amount of goods shipped from node i to node j

and recalling that

- f is the total amount shipped through the network
- f_{ij} is the given upper bound of the flow through arc (i, j)
- The conservation of flow assumption
- The nonnegativity of the flow f and the variables x_{ij}

we have the following linear-programming problem:

Maximize

$$f$$

subject to

$$
\begin{array}{rrrrrrrrrl}
x_{01} & + x_{02} & + x_{03} & & & & & & & = f \\
-x_{01} & & & + x_{12} & + x_{13} & + x_{1m} & & & & = 0 \\
 & -x_{02} & & - x_{12} & & & + x_{23} & + x_{2m} & & = 0 \\
 & & -x_{03} & & - x_{13} & & - x_{23} & & + x_{3m} & = 0 \\
 & & & & & -x_{1m} & & - x_{2m} & - x_{3m} & = -f \\
x_{01} & & & & & & & & & \leq 2 \\
 & x_{02} & & & & & & & & \leq 3 \\
 & & x_{03} & & & & & & & \leq 1 \\
 & & & x_{12} & & & & & & \leq 4 \\
 & & & & x_{13} & & & & & \leq 1 \\
 & & & & & x_{1m} & & & & \leq 3 \\
 & & & & & & x_{23} & & & \leq 1 \\
 & & & & & & & x_{2m} & & \leq 2 \\
 & & & & & & & & x_{3m} & \leq 2
\end{array}
$$

and the $x_{ij} \geq 0$ and $f \geq 0$.

The first equation states that the total amount f flowing into node 0 must equal the total amount flowing out of node 0 to the other nodes. Similar interpretations can be given for the other equations. There is one equation for each node. The inequalities represent the upper bound conditions on the flows through each arc.

The problem is certainly feasible because all of the constraints are satisfied if all $x_{ij} = 0$ and $f = 0$. But this, as they say, is a trivial solution.

Although the maximal flow problem can be solved with the aid of the simplex method, the mathematical structure of such problems lends itself to solution procedures which are faster and rather easy to implement. Quite large problems, with thousands of nodes and arcs, can be handled. We shall not, however, delve into the rudiments of these solution procedures, but we shall illustrate a few related points.

For our simple network, the maximum flow $f = 6$. On the following network for this problem, we have added a number couple, e.g., (0, 4), to each arc (i, j), where the first number represents the flow through the arc, x_{ij}, and the second number the given upper bound for that arc, f_{ij}. Of course, $x_{ij} \leq f_{ij}$. For the optimum solution, we then have

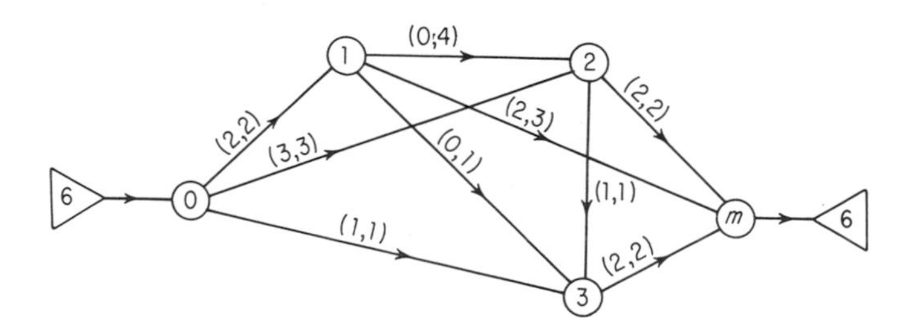

The reader will note that the arcs leaving node 0 are saturated—we cannot push any more through them. If we remove those arcs from the network—arcs (0, 1), (0, 2), and (0, 3)—we end up with the source node disconnected from the sink node. Such a set of arcs is called a "cut"—a cut being a set of arcs which if removed from the network disconnects the network into two parts, with the source in one part and the sink in the other. The arcs in a cut have a certain total capacity. Here it is 6. For the cut consisting of arcs (0, 3), (1, 3), (1, m), (2, 3), (2, m), the total arc capacity is 8. The main theorem of networks—the Max-flow Min-cut theorem—states: for any network, the maximal flow value from node 0 to node m is equal to the minimal cut capacity of all cuts separating 0 and m. A moment's reflection will cause one to say that this theorem is intuitively obvious. An algebraic proof takes a little longer. The efficient computational schemes for the maximal flow problem are based on the results of this theorem. These procedures also ensure that the values of the variables—the total flow and the flow through each arc—are given in terms of integers.

Minimal Cost Network Flow Problem

Another set of significant network problems is the class of minimal cost network flow problems. As we shall see, this type of problem includes, among others, the transportation problem and the shortest route problem.

For the minimal cost flow problem we are given, as in the maximal flow problem, a general network over which units of a homogeneous commodity are to be shipped from the source to the sink. Associated with each arc (i, j) is a cost c_{ij} of shipping one unit of the commodity from node i to node j. We must ship a given quantity of F units from the source to the sink so as to minimize the total cost of shipping the F units. We assume conservation of flow at the nodes and that the flow x_{ij} along any arc (i, j) is nonnegative and is bounded, i.e., $0 \leq x_{ij} \leq f_{ij}$. The mathematical model for this problem is quite similar to the maximal flow model, and we shall illustrate it for the network of the previous example. We must remember, however, that for our present problem we know what the total flow F is—in the maximum flow problem the total flow f was to be determined.

For our example let $F = 5$ (it cannot be greater than 6) and let the cost of shipping a unit of flow along an arc, the c_{ij}, and the upper bound for the arc, the f_{ij}, be shown on the network by the number couple $[f_{ij}, c_{ij}]$. For our network we then have

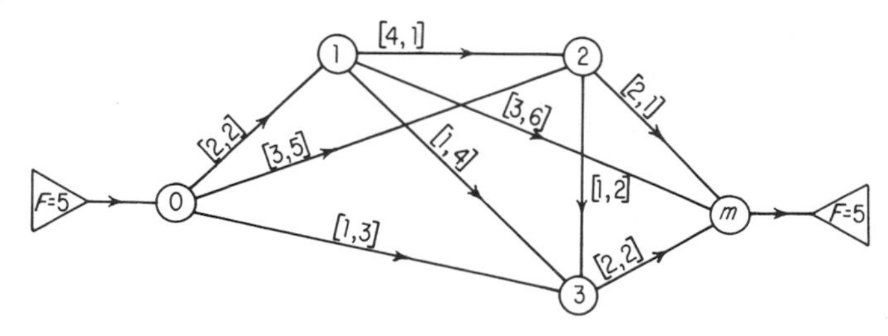

Thus, for arc (2, 3), $f_{23} = 1$ and $c_{23} = 2$.

The mathematical model is to

minimize

$$2x_{01} + 5x_{02} + 3x_{03} + x_{12} + 4x_{13} + 6x_{1m} + 2x_{23} + 2x_{2m} + 2x_{3m}$$

subject to

$$\begin{array}{lllllllllr}
x_{01} & + x_{02} & + x_{03} & & & & & & & = 5 \\
-x_{01} & & & + x_{12} & + x_{13} & + x_{1m} & & & & = 0 \\
 & -x_{02} & & - x_{12} & & & + x_{23} & + x_{2m} & & = 0 \\
 & & -x_{03} & & -x_{13} & & -x_{23} & & +x_{3m} & = 0 \\
 & & & & & -x_{1m} & & - x_{2m} & - x_{3m} & = -5 \\
x_{01} & & & & & & & & & \leq 2 \\
 & x_{02} & & & & & & & & \leq 3 \\
 & & x_{03} & & & & & & & \leq 1 \\
 & & & x_{12} & & & & & & \leq 4 \\
 & & & & x_{13} & & & & & \leq 1 \\
 & & & & & x_{1m} & & & & \leq 3 \\
 & & & & & & x_{23} & & & \leq 1 \\
 & & & & & & & x_{2m} & & \leq 2 \\
 & & & & & & & & x_{3m} & \leq 2
\end{array}$$

and all $x_{ij} \geq 0$.

We shall not describe the special procedure for solving such problems—the out-of-kilter method—except to say that it has been used to solve quite large problems in a most efficient manner. We can, however, specialize the above problem to a most interesting one, one which as a rather simple solution procedure.

The Shortest-route Problem

If we let the network be a roadmap with the source being the origin city, the sink the destination city, and the c_{ij} being the distance between cities (the intermediate nodes), we can convert the above problem to that of finding the minimum distance between the origin and the destination by letting the amount to be shipped $F = 1$ and setting all the upper bounds $f_{ij} = 1$. We are then allowing for the shipping—the flow—of one unit from the origin to the destination so that the total distance traveled by the one unit is a minimum. This problem can also, of course, be solved by the regular techniques of linear programming, but more efficient algorithms exist. We shall illustrate one for the short-route problem below.

We are given the four-city network with the distances as shown on each arc and in the distance table

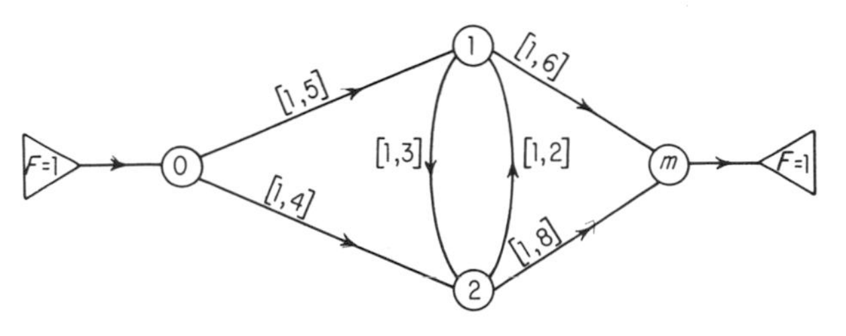

From	To		
	1	2	m
0	5	4	—
1	—	3	6
2	2	—	8

The computational procedure for this problem is a simple combinatorial node labeling scheme and consists of two basic steps.

Shortest-route Algorithm

STEP 1: Assign all nodes a label of the form $\{-, d_i\}$, where the first component indicates the preceding node in the shortest route and d_i indicates the shortest distance from node 0 to node i. Node 0 starts with a label $d_0 = 0$, with its first component always a $-$; all other nodes start with $d_i = \infty$, any very large number, and a first component of $-$.

STEP 2: For any arc (i, j) for which $d_i + c_{ij} < d_j$, change the label of node j to $\{i, d_i + c_{ij}\}$, and continue the process until no such arc can be found. In the former situation we have determined a shorter route from node 0 to node j which goes through node i; in the latter situation the process is terminated, and the node labels indicate the shortest distance from node 0 to node j.

For our problem, we have the shortest route from node 0 to node m of length 11, with the route going from node 0 through node 1 to node m. The appropriate labels are shown on the following network

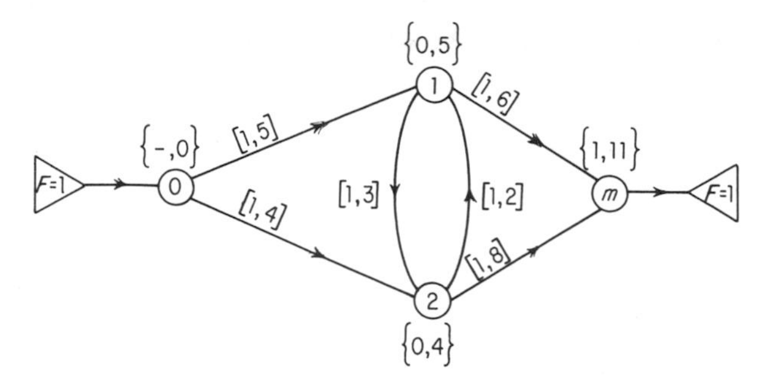

The transportation problem described in Chapter 1 can be readily interpreted as a minimal cost network flow problem. By referring to Figure 5 of that chapter, the reader should be able to add source and sink nodes in a fashion which almost satisfies the requirements of the minimal cost network flow problem. The source node has $F = 25$ refrigerators that must be shipped, 11 going to factory one and 14 to factory two. The sink node must receive 25 refrigerators, 10 from store one, 8 from store two, and 7 from store three. These conditions can be made a part of the mathematical model by constraints which force shipments of 11 and 14 from the source to the respective factories, and shipments to the sink to be exactly 10, 8, 7 from the respective stores. These types of conditions add little extra burden to the solution process of the out-of-kilter algorithm and can be included for the general minimal cost network flow problem.

THE TRAVELING-SALESMAN PROBLEM

There once was a farmer's daughter . . .
J. MILLER

The salesman's cry—"You got to know the territory!"—usually refers to his knowing where his prospects are and what goods and styles are preferred by his prospects and regular customers. For lo these many years, our salesman in question, Willie Simon of the Simple Furniture Company, has traveled about his wide territory, getting to know the lay of the land and building up an impressive sales record. At the same time, however, he was building up an impressive expense account. As part of the continuing management-consulting activities of Super Management Consultants, the SMC boys decided to take a look at the happenings in Willie's territory. Soon Willie was getting harassed by memos and questionnaires wanting to know such things as, "What happened in Boston, Willie, on 6/30? Expense account way out of line for that date." An SMC analyst joined Willie in his travels and took copious notes and data. Willie tried to make friends with the analyst in order to influence the report. When Willie learned that his companion was a mathematician, he dug deep into his salesman's repertoire of stories and attempted to regale him with such tales as the following.

Horse Sense

A mathematics professor decided to take his sabbatical by resting in the country. A local farmer, learning that his new neighbor was a mathematician, approached him with a story about the farmer's horse who could add, subtract, multiply, and divide. The professor decided to humor the farmer and paid the horse a visit. The horse could do all that was claimed. He answered correctly a number of simple problems by either stomping on the stable floor or picking up a piece of chalk in his mouth and writing on a blackboard. The professor decided to work with the horse, and before long, he had the animal doing basic algebra and Euclidean geometry.

When it was time to return to the university, the professor took the horse with him and enrolled him as a special mathematics student. During the first semester, the horse managed trigonometry without any trouble. He was getting a straight A in the second semester until the lectures turned to analytic geometry. The poor horse floundered. He just couldn't make it. He was given special tutors, but nothing helped.

The mathematics department, not wanting to lose its prize student, called a special meeting of the faculty to see what could be done. They brought in a specialist in animal husbandry, psychiatrists, jockeys, but came up with no answers. They reviewed his record—excellent in arithmetic, algebra, Euclidean geometry, trig—but analytic geometry appeared to be his Waterloo.

Finally, after much contemplation, a bright assistant instructor resolved the dilemma. "We should have guessed it," he said. "In trying to teach him analytic geometry, we were putting Descartes before the horse."

Willie's tales received no reaction. He fumed.

On his next visit to the home office, Willie had it out with his brother, President Simon, and threatened to quit unless his mode of operation and expense account were left alone. President Simon calmed Willie down and assured him that Super Management Consultants had given Willie a clean bill of health—except for one thing. The SMC study claimed that Willie was spending too much time and money in traveling from one city to the next. It appeared as if Willie wasn't arranging the visits to the successive cities in his territory to make the total distance traveled the shortest possible.

"All we want you to do, Willie," said President Simon," is to visit the cities in the order recommended by the consultants. The rest of the expense account is yours."

Much relieved, Willie agreed to put the suggested routing into effect as soon as he received it—he is still waiting. It appears as if Willie has had the last (horse) laugh on the consultants. For it turned out that, although the traveling-salesman problem can be formulated as a special linear-programming problem, the size of Willie's territory—he covers 100 cities, towns, and hamlets—made it computationally intractable. Just what was the trouble?

Let us be a little more precise and state the traveling-salesman problem as follows:

A salesman is required to visit each city of his territory. He leaves from a home city, visits each of the other cities exactly once, and finally returns to the home city. We are required to find an itinerary which minimizes the total distance traveled by the salesman.

There are a number of approaches to setting up this problem as a linear-programming problem, the mathematics of which are too involved to develop here. These formulations require that the variables be restricted to integers, either 0 or 1, and a large number of equations. For Willie's problem, one formulation would require 10,000 equations.

A look at a small, five-city problem illustrates some of the difficulties. We shall number the cities 1, 2, 3, 4, and 5, with city 1 being the home city. We know the distances between each city, and we assume a city can be reached from any other city—that is, you can get there from here. To make the discussion more general, we also assume that the distance in going from say city 2 to city 5 is not necessarily the same distance if we reversed the trip and were going from city 5 to city 2. This nonsymmetric situation could be due to one-way bypasses, detours, etc.

For our five cities we have the following distance table in miles:

City	1	2	3	4	5
1	0	17	10	15	17
2	18	0	6	12	20
3	12	5	0	14	19
4	12	11	15	0	7
5	16	21	18	6	0

For example, distance from city 1 to city 5, $d_{15} = 17$, while the distance from city 5 to city 1, $d_{51} = 16$.

As the salesman leaves his home city 1, he has a choice of going to any one of the remaining four cities—let us assume he goes to city 3. From city 3 he can travel to one of the remaining three—2, 4, or 5—and he selects 2. His next selection is city 5, then, of course city 4, and return to city 1.

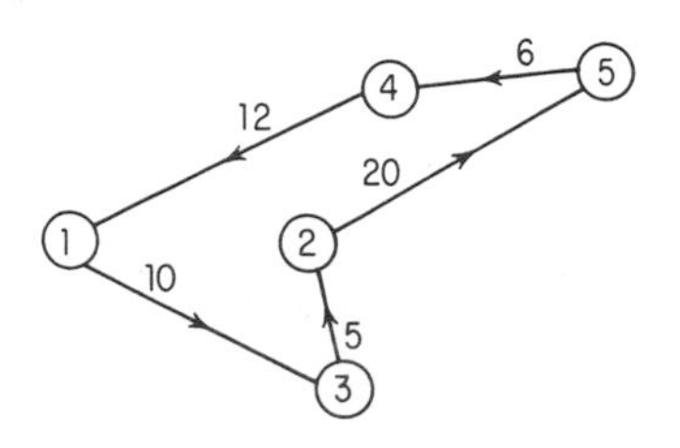

The ordering of the cities—1, 3, 2, 5, 4, 1—is called a "tour," with a tour having a corresponding tour distance. For our tour, the total distance is

$$\begin{aligned} d_{13} + d_{32} + d_{25} + d_{54} + d_{41} &= 10 + 5 + 20 + 6 + 12 \\ &= 53 \text{ miles} \end{aligned}$$

The five-city problem has a total of $4 \times 3 \times 2 \times 1 = 24$ possible tours. For such small problems, we could enumerate all tours and their distances and select the one with the smallest number of miles. However, this approach soon becomes impractical as the number of cities slowly increases. For 9 cities there are 362,880 tours, while for 14 cities there are 87,178,291,200 tours.

The linear-programming approaches to this problem look quite a bit like the personnel-assignment formulation. Here we wish to assign to a tour a link connecting a city to another city, with each city being allowed the assignment of only one link emanating from it (a person can be assigned to one job, but must be assigned). The links from city one are denoted by x_{12}, x_{13}, x_{14}, x_{15}, and for a link to be assigned we have the equation

$$x_{12} + x_{13} + x_{14} + x_{15} = 1$$

A zero value of the variable means the corresponding link is not used; a value of one means that it is used. For our sample tour the assignment table is

City	1	2	3	4	5
1	0	0	1	0	0
2	0	0	0	0	1
3	0	1	0	0	0
4	1	0	0	0	0
5	0	0	0	1	0

The difficulty in treating the salesman problem like an assignment problem is that the optimum assignment, which minimizes the total distance of the assignments, might not be a tour. The following assignment table has a distance of 49 miles, less than the sample tour's distance of 53 miles, but the assignment yields two subtours, as illustrated.

City	1	2	3	4	5
1	0	0	0	1	0
2	0	0	1	0	0
3	0	1	0	0	0
4	0	0	0	0	1
5	1	0	0	0	0

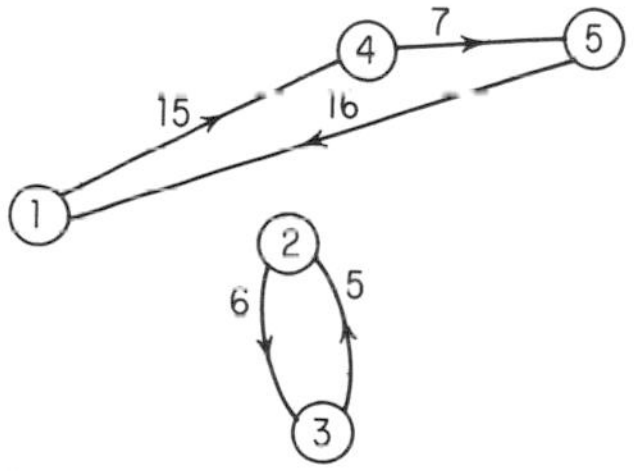

The assignment approach must be augmented by conditions which rule out the generation of assignments which are constructed of such subtours. This is the reason why the linear-programming approach requires so many equations.

Willie Simon's satisfaction will, however, be short-lived. There are other approaches to solving the traveling-salesman problem, although the largest problem solved has had only 70 cities. The procedure of dynamic programming appears good for small problems up to 13 cities; the scheme called "branch-and-bound" solved the 70-city problem; while integer-programming procedures have been used to solve a 42-city problem. Willie's problem could be broken up into smaller problems in that cities close to one another could be treated as one city. But, as in all such cases, it is up to the analysts, armed with their algorithms and computers, to prove that their mathematical solutions are better and can be put to work.

We shall leave it to the reader to determine if the sample tour—1, 3, 2, 5, 4, 1—is the minimum distance tour, and if not, which tour is the minimum. The concept of the traveling-salesman problem can be extended to other fields—mail pickup and delivery, school bus scheduling. We should not be reticent to tackle such problems due to the apparent computational difficulties. Some level of improvement is usually attainable, and there is always the chance of hearing a good story.

THE CONTRACT-AWARDS PROBLEM

Sold!
AUCTIONEER'S CRY

Whenever we are faced with an optimization problem, our natural inclination is to make sure we select a feasible solution which is also an optimum solution. We want to do our best no matter what—or do we? As the computational procedures of linear programming search for the optimum by first finding a solution, then a better one, then an even better one, and so on, we might question the need to actually wring out of our computational washing machine the last penny of savings. The labor and computational costs to do so might cost more than the savings afforded by improving upon the solution at hand. Do we really want the minimum cost diet or just a good diet which satisfies the daily requirements? Can we live with Sid Simon's solutions to his activity-analysis problems, or do we really want to make the last dollar?

For some problems a good solution would suffice, but in general, if the economics of the computation are favorable, the optimum solution should be found and used. There are some problems, however, where the use of an optimum solution is a must. In particular, with the contract-awards problem the need for determining the lowest cost solution is dictated by the U.S. Congress.

The Armed Forces Procurement Act requires the Armed Forces to purchase supplies on a competetive basis. Contracts must be advertised, sealed bids submitted, and the awards made so as to minimize the total cost to the government. Once the contracts have been awarded, the procurement office must be ready to demonstrate—especially to the satisfaction of the losing bidders and their Congressmen—that the total cost was the least possible.

The award of the contracts can be a rather complicated decision-making problem as constraints imposed by the bidders can be quite complex. As all the bids are made public after an award is made, the contract evaluators have to be very astute in juggling and analyzing the data. With the advent of linear programming however, this

heavy burden—and some ulcers—have been removed. How does linear programming fit in? Let's take a closer look at the problem.

For a given procurement, the individual manufacturer submits bids on which he states

- The price per unit of article or articles

- The maximum or minimum quantity of each item that can be produced at the stated price
- Any other conditions he wishes to impose

The bid reflects the manufacturer's desire for profit, his guess about the other fellow's bid, and his own peculiar limitations.

In evaluating the bids, the procurement office must add shipping and other related costs to each bidder's quoted prices. Similarly, any savings that could be effected by agreeing to certain conditions are subtracted; e.g., a discount may be allowed for payments made within a certain time. The basic contract-awards problem can be formulated as a transportation problem, but certain conditions imposed by the bidders require special handling by the problem formulator. For example, a bidder might require his manufacturing at least 500 units and no more than 1,000 units; or he might have different prices for each lot of 1,000 units to reflect quantity discounts. The following example is an actual contract-awards problem as encountered by an army purchasing agency. We shall only present the facts of the case and leave it to the reader to develop the equations and approach to solving it.

The quartermaster depots listed below require the stated amount of packages of duplicating machine stencils. The amounts required are separated into domestic and export categories as the special packaging for export dictates a higher cost per unit, in Table 1.

Four manufacturers submit bids. Bidder 1 requires an award of at least 11,500 packages and allows for a maximum award of

TABLE 1

Depots	Requirements	
	Domestic	Export
Columbus	10,395	
Richmond	12,420	
San Antonio	10,395	
Schenectady	9,720	39,555
Utah	3,240	
Sharpe	5,535	
Auburn	3,645	
Atlanta	10,330	3,510
Totals	65,680	43,065
Grand total	108,745	

33,145 packages for domestic and a maximum of 3,510 for export. Bidder 2 has no minimum or maximum conditions. Bidder 3 offers a maximum of 60,000 packages, while bidder 4 bids only for the domestic contract and has no other conditions. This information, along with the cost per package for each bidder–depot combination, is summarized in Table 2. The given prices include packaging and shipping costs; e.g., it costs $.6868 to ship one package from bidder 2 to the Columbus depot.

TABLE 2 Costs and Amounts Bid

City	No. required	Maximum no. offered				
		Bidder 1 Domestic	Bidder 1 Export	Bidder 2	Bidder 3	Bidder 4
		33,145	3,510	108,745	60,000	65,680
Columbus	10,395	0.7289	—	0.6868	0.6574	0.6832
Richmond	12,420	0.7398	—	0.7058	0.6489	0.6724
San Antonio	10,395	0.7229	—	0.7204	0.6904	0.7227
Schenectady domestic	9,720	0.7406	—	0.7075	0.6318	0.6627
Schenectady export	39,555	—	0.7749	0.7319	0.6452	—
Utah	3,240	0.7247	—	0.7358	0.6944	0.7306
Sharpe	5,535	0.7276	—	0.7389	0.6973	0.7339
Auburn	3,645	0.7297	—	0.7389	0.6973	0.7339
Atlanta domestic	10,330	0.7325	—	0.7049	0.6646	0.6917
Atlanta export	3,510	—	0.7663	0.7291	0.6816	—

The solution which satisfies the conditions imposed by the bidders and minimizes the total cost to the purchasing agency is given in Table 3. This award results in a total cost to the government of $72,953.1935. The linear-programming procedures which yield this solution guarantee that no better result is possible.

Historical Footnote

The relationship between the contract-awards problem and the transportation problem was first discovered and exploited by mathematicians of the U.S. National Bureau of Standards (NBS), working with

TABLE 3 Contract Award Minimum Solution

City	Bidder 1	Bidder 2	Bidder 3	Bidder 4
Columbus	—	—	—	$10,395
Richmond	—	—	—	$12,420
San Antonio	—	$10,395	—	—
Schenectady domestic	—	—	$4,515	$5,205
Schenectady export	—	—	$39,555	—
Utah	—	—	$3,240	—
Sharpe	—	—	$5,535	—
Auburn	—	—	$3,645	—
Atlanta domestic	—	—	—	$10,330
Atlanta export	—	—	$3,510	—

personnel from the Philadelphia quartermaster depot. Prior to the advent of linear programming, each problem was solved by submitting the bids to a series of evaluations conducted by different analysts. When no change could be found, the successful bidders would be announced, and everyone would hope for the best. When the first operational tests were conducted using the NBS computer, the SEAC, the quartermaster group continued to solve the problems with their analysts in order to build up the necessary confidence in the linear-programming approach. The computer solutions were always better than, or at least as good as, the quartermaster solutions. In fact, this test again demonstrated that experienced personnel can do a good job in finding the optimum solution, if the data and constraints are not too involved. However, as many such problems have to be solved on a daily, routine basis, the computerized linear-programming approach has other economic advantages, along with its ability to determine the low-cost solution.

THE THEORY OF GAMES

The play's the thing.
SHAKESPEARE

Like linear programming, the theory of games can be considered as a modern development in the field of mathematics. To the casual observer this would appear to be the only element which these areas have in common. For whereas in the general linear-programming problem, we are concerned with the efficient use or allocation of limited resources to meet desired objectives, in the theory of games we are interested in developing a pattern or strategy of play for a given game which will enable us to win as much as possible. A remarkable correspondence between these problems exists in that the mathematical model of an important class of game-theory problems is identical to a linear-programming model. We shall describe some of the basic concepts of game theory and delve into its relationship to linear programming.

In general, the main concern of the theory of games is the study of the following problem: if n players, denoted by $P_1, P_2, \ldots, P_n$ play a given game, how must each player play to achieve the most favorable result? In interpreting this basic problem, we say that the term "game" refers to a set of rules and conventions for playing and a "play" refers to a particular possible realization of the rules, i.e., an individual contest. At the end of a play of a game, each of the players receives an amount of money, called the "payoff." For those players that lose, the payoff is a flow of cash out of his pocket into the pockets of the winning players. If all the money which is lost and won stays with the players, the game is called "zero-sum." A game which is not zero-sum, for example, would be poker in which the house takes a percentage of each pot.

Games are also classified by the number of players and possible moves. Chess is a two-person game with a finite number of moves (if we include appropriate "stop rules"), and poker is a many-person game, also with a finite number of moves. A duel in which the duelists may fire at any instant in a given time interval is a two-person game with an infinite number of possible moves. Games are further characterized by being cooperative or noncooperative. In the former, the players have the ability to gang up on other players and work as teams, while in the latter each player is concerned only with his own result. Two-person games are, of course, noncooperative. We shall only discuss finite, zero-sum, two-person games in that it is this type of game which bears a close relationship to linear programming.

As our first example of such a game, let us consider the now classical analysis of a real-life strategic situation—the Battle of the Bismark Sea.[1]

[1] O. G. Haywood, Jr., "Military Decision and Game Theory," *Journal of The Operations Research Society of America*, vol. 2, no. 4, 1954.

How the Battle Got Its Name

In February–March, 1943 intelligence reports indicated that a Japanese troop-and-supply convoy was assembling at Rabaul, New Britain, for movement to Lae, New Guinea. General Kenney, the Commander of the Allied Air Forces in the Southwest Pacific Area, was ordered by General MacArthur to intercept and inflict maximum destruction on the convoy.

The Japanese Commander—whose name does not appear to have been recorded—had the choice of sending the 16-ship convoy north of New Britain or south of that island. Either route required three days. These choices represented his possible strategies. Weather reports—which were available to both sides—indicated rain and poor visibility over the northern route, while it was expected to be clear in the south.

General Kenney limited himself to two possible courses of action. He could concentrate most of his search aircraft on one route or the other. The bombing force could strike the convoy on either route soon after it was found.

General Kenney's mission was to find the convoy and cause it to suffer as many days exposure to his bombers as possible; the Japanese commander desired the minimum exposure to bombing. Each would make his decision independent of the other. What course of action should each choose? We first need to develop some additional information before the strategies can be selected.

Taking the weather, mobility of forces, and related constraints into consideration, General Kenney could construct a 2×2 tableau —or as we shall call it a "payoff matrix"—which indicated how many days of bombing he would have relative to how he deployed his search aircraft. The analysis goes something like this.

1. Kenney strategy—concentrate search aircraft in north
   a. Convoy sails north
      Weather hampers search, but can get two days bombing because of proper concentration of search aircraft.
   b. Convoy sails south
      Convoy in clear weather, but because of limited search aircraft can only expect two days of bombing.
2. Kenney strategy—concentrate search aircraft in south
   a. Convoy sails north
      Poor visibility and badly positioned search aircraft limits bombing to one day.
   b. Convoy sails south
      All things going properly—good weather, bulk of search aircraft—convoy would suffer three days of bombing.

Arranging this data in payoff matrix form, we have

		Japanese strategies (convoy route direction)	
		Northern route	Southern route
Kenney strategies (search aircraft concentration)	Northern route	2 days	2 days
	Southern route	1 day	3 days

To treat this military situation as a problem in game theory, we have to assume that the Japanese commander would have the same knowledge as General Kenney—the Japanese commander should be able to construct the same payoff matrix and interpret it in a similar fashion. The game is zero-sum in that a day's bombing gained by General Kenney is a day lost by the Japanese; i.e., an outcome judged good by one commander is judged equally bad by the other. Part of the difficulty with game theory is obtaining the agreement between the antagonists as to the actual utility of the numbers that make up the payoff matrix. Such discussions would be discursive to our presentation and are better left to the treatises on game theory.

By analyzing this simple two-strategy problem many of the features of game theory materialize. Let us generalize the concept of a payoff matrix to the concise notation of matrix algebra and represent the game as a 2×2 matrix, with the rows corresponding to the strategies of the maximizing player—here General Kenney—and the columns to the minimizing player—the Japanese commander. We have

$$\begin{pmatrix} 2 & 2 \\ 1 & 3 \end{pmatrix}$$

as our payoff matrix.

We now join the commanders sitting in their respective headquarters staring at the numbers, trying to determine what to do. The Japanese commander, an expert in the game of Go, feels confident that he can select the best strategy for his side. The numbers in the payoff matrix cause him some concern, since it appears as if no matter what he does, his convoy will be found and subject to some damage. It is a no-win situation. The best he can hope for is one day's exposure to the bombers. Can he achieve that goal?

In comparing the columns—the northern-strategy payoffs to the southern-strategy payoffs—the Japanese commander notices that he should not select the southern strategy. If he did, he could expose himself to two days or possibly three days of bombing, while the northern-convoy route's corresponding exposure was two days or one day. The Japanese commander crosses that strategy from his matrix. He can even send such information direct to General Kenney, for the General, being a master strategist, would also remove the same column from his matrix.

The General now has a reduced matrix of two rows, representing his strategies, and one column, representing the fact that the Japanese must take the northern route. The reduced payoff matrix is

$$\begin{pmatrix} 2 \\ 1 \end{pmatrix}$$

The problem is now quite simple. As General Kenney wishes to maximize the days of bombing he selects the northern route. And, as though calculated by the mathematicians, the Japanese actually took the northern route, General Kenney actually sent the bulk of his search aircraft to the northern route, and the Battle of the Bismark Sea occurred in the right place.

The solution to this strategic problem is of a special kind—we say it is solved by using a pure strategy—here the northern route is selected by each commander, positively. A commander or player doesn't even consider the other available strategy—he doesn't throw dice or a coin to make his decision. The reason for this is that the matrix of the Bismark Sea problem contains what is called a "saddle point." For a matrix, a saddle point is an element which is both the minimum of its row and the maximum of its column—two days of bombing at the intersection of the northern-route strategies is such a number. The corresponding row and column represent the optimum strategies to be used by the opposing player, and the payoff value is the value of the saddle point. The reason for this is the underlying conservatism inherent in the theory of games. Let us explain.

We assume the following (3×3) matrix is associated with some game

$$\begin{pmatrix} 3 & 5 & 6 \\ 2 & -1 & 3 \\ 0 & 7 & 4 \end{pmatrix}$$

Recall that the numbers in the payoff matrix are written in terms of the maximizing player (e.g., General Kenney), so a positive number means the row player wins, a negative number means he loses that amount, while a zero means no money exchanges hands. The row player, player one as he is called, looks at the numbers in his first row and notes that the worst thing that could happen to him if he selected the first strategy is that the column player, player two, also selects his first strategy. If this happens, player one receives three units instead of five or six units. Player one performs the same analysis on the other two rows and finds he would gain -1 unit, i.e., lose one unit, and 0 units, respectively. We write these minimum row numbers along side the matrix

$$\begin{pmatrix} 3 & 5 & 6 \\ 2 & -1 & 3 \\ 0 & 7 & 4 \end{pmatrix} \begin{matrix} \textcircled{3} \\ -1 \\ 0 \end{matrix}$$

The added numbers represent the worst things that could happen to player one for each one of his strategies. Player one can thus select the best of these worst events and can decide to always play strategy 1—here he will gain at least three units for every play of the game. Can player two do anything to reduce his loss per play?

Performing a similar analysis in terms of player two's frame of reference, we see that if he selects the first column, the worst thing that could happen to him is player one selects the first row; i.e., player two would have to pay player one a total of three units. For columns 2 and 3 the payoffs would be 7 and 6, respectively. Adjoining these numbers to the matrix we have

$$\begin{matrix} \begin{pmatrix} 3 & 5 & 6 \\ 2 & -1 & 3 \\ 0 & 7 & 4 \end{pmatrix} & \begin{matrix} \textcircled{3} \\ -1 \\ 0 \end{matrix} \\ \begin{matrix} \textcircled{3} & 7 & 6 \end{matrix} & \end{matrix}$$

The best thing player two can do in this situation, besides not play the game, is to always play the first column and always lose three units. Thus, the element 3 in the matrix is a saddle point. Games with saddle points are readily solved by the above analysis. We see that if a player deviates from his pure strategy which corresponds to the location (row or column) of a saddle point, he leaves himself open to a greater loss or less winnings.

The interesting games, however, are those without saddle points. If we attempt the saddle-point analysis on the game defined by the (3×4) matrix

$$\begin{pmatrix} 1 & 5 & 0 & 4 \\ 2 & 1 & 3 & 3 \\ 4 & 2 & -1 & 0 \end{pmatrix}$$

we see that the best of the worst things—this is called the "max-min"—for player one occurs for the second row, and the worst of the best things—the "min-max"—for player two occurs for the third column

$$\begin{matrix} \begin{pmatrix} 1 & 5 & 0 & 4 \\ 2 & 1 & 3 & 3 \\ 4 & 2 & -1 & 0 \end{pmatrix} & \begin{matrix} 0 \\ \textcircled{1} \\ -1 \end{matrix} \\ \begin{matrix} 4 & \;\; 5 & \;\; \textcircled{3} & \;\; 4 \end{matrix} & \end{matrix}$$

Here we do not have a saddle point However, player one can guarantee winning at least 1 unit by always playing his pure max-min strategy 2; while player two can guarantee not losing more than three units by always playing his pure min-max strategy 3. For games of this sort, the players can gain a better payoff, on the average, if they allow themselves to select among the available strategies in a probabilistic fashion. Here, player one should be able to win more than one unit, while player two's losses should be less than three units. By allowing themselves to mix the strategies and randomly select a particular one for a particular play of the game, the players can drive the expected value of the game to a number between the max-min value of 1 unit and the min-max value of 3 units; i.e., each does better than the corresponding pure max-min–min-max strategies. We shall illustrate this concept, the use of a mixed strategy, with the well-known game of matching pennies.

Heads I Win . . .

To accomplish one move in this game, which here is equivalent to a play, the first player selects either heads or tails, and the second player, not knowing the other's choice, also selects heads or tails. After the choices are made known to each player, player two pays player one one unit $+1$ if they match or he receives one unit -1 if they do not. The payoff of -1 represents the giving of one unit by player one to player two. The above can be summarized by the matrix:

		Player Two Selections	
		Heads	Tails
Player One Selections	Heads	1	−1
	Tails	−1	1

We see that the matrix has no saddle point—the max-min is $+1$ and the min-max is -1.

We next define mixed strategies for each of the players. Player one would like to randomize the selection of heads or tails in that, if he used a fixed rule like "always play heads," his opponent, being a rational player (an assumption of game theory), would take advantage of such an aberration. We let x_1 be the probability that player one will select heads and x_2 the probability he will select tails. We similarly define y_1 and y_2 for player two. A typical set of probabilities would be that player one selects his first strategy with probability $3/4$ and the second with probability $1/4$. The probabilities are interpreted in terms of the frequency of playing each strategy. For the mixed strategy $x_1 = 3/4$ and $x_2 = 1/4$, as player one played the game over and over again, he would find himself, on the average, selecting the first strategy three-fourths of the time and the second strategy one-fourth of the time. As the probabilities for each player must sum to one, we always have $x_1 + x_2 = 1$ and $y_1 + y_2 = 1$, with the x's and y's nonnegative.

By definition, a solution to a zero-sum game is a pair of optimal mixed strategies—one for each player—and a number v, the value of the game, such that if player one uses his optimal mixed strategy against any strategy of player two, player one will be assured of winning at least v; and if player two plays his optimal mixed strategy against any strategy of player one, player two can be assured of not losing more than v. The main theorem of game theory shows that such optimal strategies and a number v always exists for any matrix.

We would next like to translate the above to a mathematical model for the matching-pennies problem without, as they say, making a federal case out of the underlying mathematical detail. The reader must then bear with us or read very fast.

Player one wishes to maximize the value of the game v subject to the condition that his probabilities $x_1 + x_2 = 1$ and his expected winnings, if he plays the strategy (x_1, x_2) against each of player two's pure strategies, will be at least as great as v, that is,

$$\begin{aligned} x_1 - x_2 &\geq v \\ -x_1 + x_2 &\geq v \end{aligned}$$

The inequalities are obtained by multiplying the unknown nonnegative probabilities by the corresponding elements in the columns of the payoff matrix, as indicated for the first column by

$$\begin{matrix} x_1 \times \\ + \\ x_2 \times \end{matrix} \begin{pmatrix} 1 & -1 \\ & \\ -1 & 1 \end{pmatrix}$$

Each such product represents the mathematical expectation for player one as player two selects his corresponding pure strategy. Player one wants his expectation to be at least as great as the unknown value of the game, v. The value of v is not restricted as to its sign. A positive v means the game is biased to player one; a $v = 0$ means the game is fair, while a negative v indicates the game is biased to player two. Player one is interested in maximizing v and finding nonnegative values of x_1 and x_2 such that

$$\begin{aligned} x_1 - x_2 &\geq v \\ -x_1 + x_2 &\geq v \\ x_1 + x_2 &= 1 \end{aligned}$$

The number v is also a variable of the problem, and except for allowing v to be negative as well as positive, the above problem is essentially a linear-programming problem. (This slight deviation can be taken care of in a number of ways. We can add a large positive number w to the game matrix so that all the elements are positive. The optimal strategies will be the same, but the new value would be equal to $v + w$, a positive number.)

A similar linear-programming problem can be constructed for the second player. His problem would be to minimize v subject to

$$\begin{aligned} y_1 - y_2 &\leq v \\ -y_1 + y_2 &\leq v \\ y_1 + y_2 &= 1 \end{aligned}$$

with $y_1 \geq 0$, $y_2 \geq 0$. The value of v will be the same. For the matching-pennies game, $v = 0$; i.e., it is a fair game.

The linear-programming problems of player one and player two represent two problems which have a strong mathematical relationship in terms of primal and dual linear-programming problems. (Our game problems need a slight adjustment to be true primal and dual problems.)

By recalling the geometric approach to solving linear-programming problems described in Chapter 3, we can develop an easy way to solve game-theory problems in which each opponent has a choice of only two strategies, i.e., (2×2) games. (The procedure to be described can, in fact, be extended to solving $(2 \times m)$ games, where one player has only two strategies and the other has m strategies.) We shall solve the problem in terms of player one and leave it to the reader to solve it for player two. The problem to be considered, then, is the following:

Maximize

$$v$$

subject to

$$\begin{aligned} x_1 - x_2 &\geq v \\ -x_1 + x_2 &\geq v \\ x_1 + x_2 &= 1 \\ x_1 \qquad &\geq 0 \\ x_2 &\geq 0 \end{aligned}$$

As the geometric approach to solving such problems is restricted to those having only two variables, we must first convert this problem, which now has the three variables (x_1, x_2, v), to a problem with only two variables. This can be readily accomplished by using the equation $x_1 + x_2 = 1$ to represent one of the variables in terms of the other; i.e., we let $x_2 = 1 - x_1$ and substitute this expression of x_2 in terms of x_1 to obtain the new constraints

$$\begin{aligned} x_1 - (1 - x_1) &\geq v \\ -x_1 + (1 - x_1) &\geq v \\ 0 \leq x_1 &\leq 1 \end{aligned}$$

or

$$\begin{aligned} 2x_1 - v &\geq 1 \\ -2x_1 - v &\geq -1 \\ 0 \leq x_1 &\leq 1 \end{aligned}$$

As this is player one's game model, we want to find the maximum value of v, recalling that v, the value of the game, is unrestricted as to its sign. We construct a two-dimensional graph with x_1 for one dimension and v for the other

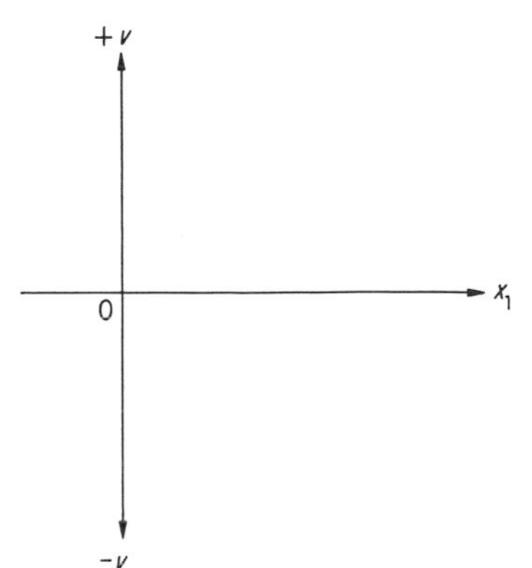

The inequality $0 \leq x_1 \leq 1$ restricts the value of x_1 to be a nonnegative number less than or equal to one, and this is represented on the graph by the unbounded shaded area

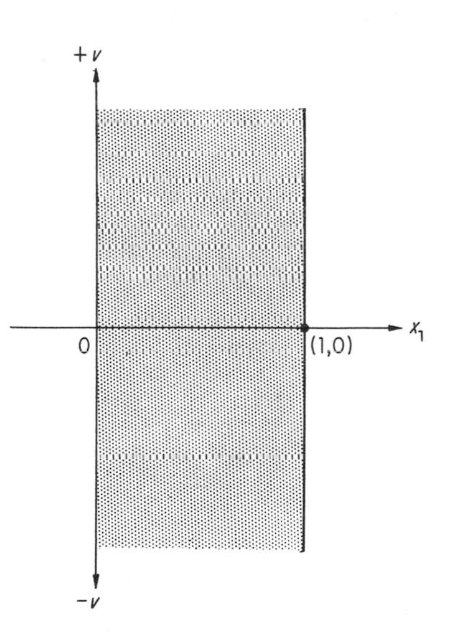

We next draw the lines $2x_1 - v = 1$ and $-2x_1 - v = -1$ to obtain the joint solution space for the corresponding inequalities. This space is represented by the shaded area

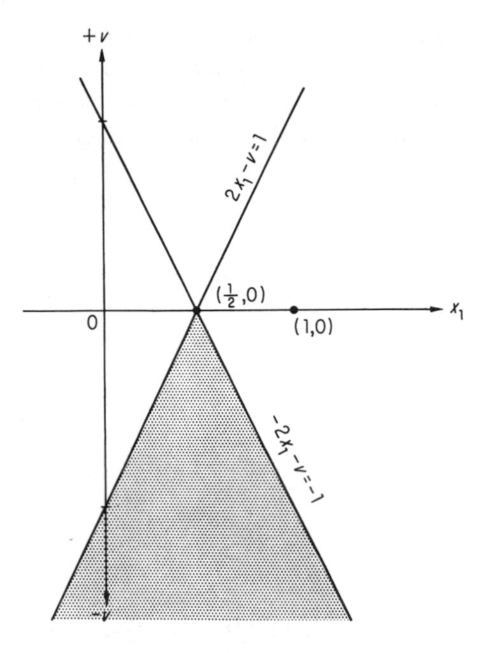

Superimposing these two graphs, we obtain the points which simultaneously satisfy the three constraints of the problem

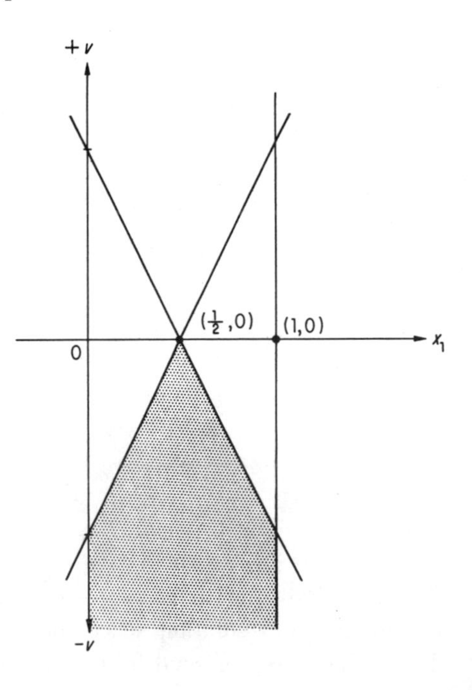

As we are looking for the point in the shaded region which has the maximum value of v, the optimum solution point is $x_1 = 1/2$, $v = 0$. The solution is unique, as all other solutions have a negative value of v. Since $x_2 = 1 - x_1$, the optimum mixed strategy for player one is to randomize the selection of his two strategies so that the frequency of selection is $1/2$ for each one; $x_1 = 1/2$, $x_2 = 1/2$. This strategy assures him an average payoff of zero; i.e., the game is fair. Player two has the same optimum strategy.

The Skin Game

For our last example, we join the brothers Simon at the annual Simple Manufacturing Company's picnic which is being held in conjunction with the county fair. Included among the guests is the ubiquitous Super Management Consultant team. Brother Si Simon has just rejoined the group after making the rounds of the bingo parlor, roulette wheel, and other games of chance. In fact, he really returned to borrow some money. Although his gaming ventures had emptied his pockets, his return route had carried him past a new game which looked rather easy to beat. It was a card game, and after studying the rules, Si felt that he had a winning strategy. But none of his brothers, who all knew Si to be a born loser (he was vice-president in charge of returned goods), would lend him any money.

In desperation, he approached the Super Management Consultant team and explained his needs. "Tell us about this new game," they inquired, "and let us analyze it for you. If our analysis bears you out, we'll take up a collection."

"I know I can beat this game," Si replied. "Here are the rules: I play against the carnival man—we each have three cards. He has an ace of diamonds, an ace of clubs, and a two of diamonds; I also have an ace of diamonds, and an ace of clubs, but my third card is a two of clubs. We each select a card from our hands and simultaneously show it to each other. I win if the suits don't match; he wins if they do. If the two deuces are shown, there is no payoff. Otherwise, the amount of the payoff is the numerical value of the card shown by the winner. That's all there is to it. Pretty easy. All I need to do is mix up how I play my clubs and maybe throw in my ace of diamonds a few times. I'm sure to win. Oh, yes, they call this the skin game for some reason or other."

The SMC boys recognized that this was a zero-sum two-person game with three possible strategies for each player. They huddled together, drawing figures on the paper picnic napkins until they reached a unanimous decision. Si should go elsewhere for his money—or better still, he should not play this game. Their analysis went like this:

The carnival man is the maximizing player one and Si the minimizing player two. The payoff matrix is

$$
\begin{array}{ll}
 & \text{Si's strategies} \\
 & \begin{array}{ccc} \diamondsuit & \clubsuit & 2\clubsuit \end{array} \\
\begin{array}{c} \text{Carnival man's} \\ \text{strategies} \end{array}
\begin{array}{c} \diamondsuit \\ \clubsuit \\ 2\diamondsuit \end{array}
 & \begin{pmatrix} 1 & -1 & -2 \\ -1 & 1 & 1 \\ 2 & -1 & 0 \end{pmatrix}
\end{array}
$$

The carnival man would never select his first strategy—show the ace of diamonds—in that he can always get as much or better if he played the two of diamonds; i.e., the payoffs for his third strategy are equal to or greater than the corresponding payoffs for his first strategy. Hence, he really plays the reduced (2×3) game

$$
\begin{array}{ll}
 & \begin{array}{ccc} \diamondsuit & \clubsuit & 2\clubsuit \end{array} \\
\begin{array}{c} \clubsuit \\ 2\diamondsuit \end{array}
 & \begin{pmatrix} -1 & 1 & 1 \\ 2 & -1 & 0 \end{pmatrix}
\end{array}
$$

For this game Si would never play his third strategy—the deuce of clubs—in that he could do just as well, if not better, if he played his second strategy. The game finally reduces to the (2×2) game

$$
\begin{array}{ll}
 & \begin{array}{cc} \diamondsuit & \clubsuit \end{array} \\
\begin{array}{c} \clubsuit \\ 2\diamondsuit \end{array}
 & \begin{pmatrix} -1 & 1 \\ 2 & -1 \end{pmatrix}
\end{array}
$$

Letting x_2 and x_3 be the probabilities that the carnival man plays his second and third strategies (here $x_1 = 0$), the model for this game is to maximize v subject to

$$
\begin{aligned}
-x_2 + 2x_3 &\geq v \\
x_2 - x_3 &\geq v \\
x_2 + x_3 &= 1 \\
x_2 &\geq 0 \\
x_3 &\geq 0
\end{aligned}
$$

To convert it to a two-variable problem we let $x_3 = 1 - x_2$ and obtain the constraints

$$\begin{aligned} -x_2 + 2(1 - x_2) &\geq v \\ x_2 - (1 - x_2) &\geq v \\ 0 \leq x_2 &\leq 1 \end{aligned}$$

or

$$\begin{aligned} -3x_2 - v &\geq -2 \\ 2x_2 - v &\geq 1 \\ 0 \leq x_2 &\leq 1 \end{aligned}$$

The solution space in terms of x_2 and v is the shaded area.

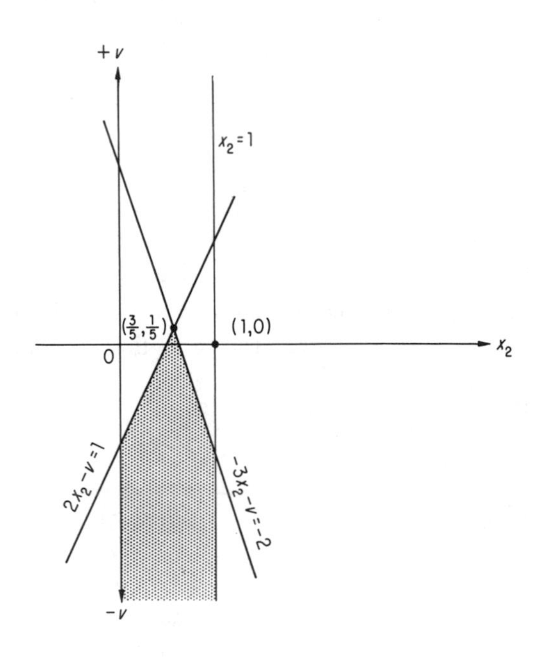

The optimum point is $x_2 = 3/5$ and $v = 1/5$. Thus, the value of the game is $1/5$ of a unit to the carnival man if he never plays his first strategy, plays the second with probability $3/5$, and plays his third strategy with probability $2/5$; i.e., $x_1 = 0$, $x_2 = 3/5$, $x_3 = 2/5$, and $v = 1/5$. The game is not fair; it is biased towards the carnival, and Si would continue his losing streak. But this logical analysis had no effect on Si, for "there's one born every minute."

MATHEMATICAL APPENDIX AND SUMMARY OF APPLICATIONS

{*Whatsoever thou takest in hand, remember the end, and thou shall never do amiss.*}

APOCRYPHA

This Appendix should enable the reader to construct a bridge between the level of presentation of this book and related works in linear programming. A safe crossing via this bridge requires an exacting toll—a toll which is coined from materials found among the diverse fields of mathematics, computing science, economics, management science, industrial engineering, and operations research. Above all, however, intuition and imaginative creativity must be alloyed with these materials in order to form a metal which will raise the tollgate.

I. INTRODUCTION[1]

Programming problems are concerned with the efficient use or allocation of limited resources to meet desired objectives. These problems are characterized by the large number of solutions that satisfy the basic conditions of each problem. The selection of a particular solution as the best solution to a problem depends on some aim or overall objective that is implied in

[1] This material is based on Garvin [28], Gass [29], Hadley [36], and Vajda [55]. Numbers in brackets refer to the publications listed in the references at the end of the Appendix.

the statement of the problem. A solution that satisfies both the conditions of the problem and the given objective is termed an "optimum solution." A typical example is that of the manufacturer who must determine what combination of his available resources will enable him to manufacture his products in a way which not only satisfies his production schedule, but also maximizes his profit. This problem has as its basic conditions the limitations of the available resources and the requirements of the production schedule, and as its objective, the desire of the manufacturer to maximize his gain.

We shall consider only a very special subclass of programming problems called linear-programming problems. A linear-programming problem differs from the general variety in that a mathematical model or description of the problem can be stated, using relationships which are called "straight-line," or linear. Mathematically, these relationships are of the form

$$a_1x_1 + a_2x_2 + \cdots + a_jx_j + \cdots + a_nx_n = a_o$$

where the a_j's are known coefficients and the x_j's are unknown variables. The mathematical statement of a linear-programming problem includes a set of simultaneous linear equations and/or inequalities which represent the conditions of the problem and a linear function which expresses the objective of the problem.

This Appendix describes the basic concepts of linear programming, reviews the standard computational techniques, and summarizes the wide variety of applications. As space does not permit the development of the theoretical and mathematical foundations of linear programming and as it is felt that one cannot take full advantage of the power of this mathematical area unless he understands its foundations, the reader is urged to do additional reading from those items listed in the bibliography.

II. LINEAR-PROGRAMMING DEFINITIONS AND BASIC THEOREMS

The *linear-programming problem* is as follows:

Find a set of numbers $x_1, x_2, \ldots, x_n$ which minimizes (or maximizes) the *linear objective function.*

$$c_1x_1 + c_2x_2 + \cdots + c_jx_j + \cdots + c_nx_n \tag{1}$$

subject to the *linear constraints*

$$\begin{array}{ccccccccc} a_{11}x_1 + & a_{12}x_2 + & \cdots & + & a_{1j}x_j + & \cdots & + & a_{1n}x_n = & a_{10} \\ \vdots & \vdots & & & \vdots & & & \vdots & \vdots \\ a_{i1}x_1 + & a_{i2}x_2 + & \cdots & + & a_{ij}x_j + & \cdots & + & a_{in}x_n = & a_{i0} \\ \vdots & \vdots & & & \vdots & & & \vdots & \vdots \\ a_{m1}x_1 + & a_{m2}x_2 + & \cdots & + & a_{mj}x_j + & \cdots & + & a_{mn}x_n = & a_{m0} \end{array} \tag{2}$$

and the *nonnegativity constraints*

$$
\begin{aligned}
x_1 \qquad\qquad & \geq 0 \\
x_2 \qquad\quad & \geq 0 \\
\ddots \qquad & \ \vdots \\
x_j \quad & \geq 0 \\
\ddots & \ \vdots \\
x_n & \geq 0
\end{aligned}
\tag{3}
$$

In words, in a linear-programming problem one wishes to find a nonnegative solution to a set of linear constraints which optimizes, i.e., minimizes or maximizes, a linear objective function. The c_j are called cost coefficients and the a_{i0} are called right-hand side coefficients. The linear constraints can be equations, as noted above, or linear inequalities of the form

$$a_{i1}x_1 + \cdots + a_{in}x_n \leq a_{i0} \tag{4}$$

or

$$a_{i1}x_1 + \cdots + a_{in}x_n \geq a_{i0} \tag{5}$$

Inequalities like (4) and (5) can be transformed to equations by the suitable addition or subtraction of a nonnegative *slack variable.* For (4) we would have

$$
\begin{aligned}
a_{i1}x_1 + \cdots + a_{in}x_n + x_{n+1} &= a_{i0} \\
x_{n+1} &\geq 0
\end{aligned}
\tag{6}
$$

and for (5) we would have

$$
\begin{aligned}
a_{i1}x_1 + \cdots + a_{in}x_n - x_{n+1} &= a_{i0} \\
x_{n+1} &\geq 0
\end{aligned}
\tag{7}
$$

Note that each inequality has a different slack variable associated with it. Each slack variable measures the difference between the left and right-hand sides of the given inequality.

As the form of a linear-programming problem requires that all variables be nonnegative, we should note that a variable not restricted to be nonnegative can always be expressed as the difference between two nonnegative variables; e.g., $x_i = x_i' - x_i''$, where $x_i' \geq 0$ and $x_i'' \geq 0$.

A linear-programming problem can contain any mixture of linear constraints. For computational purposes, the basic constraints (2) of a linear-programming problem must always be expressed in terms of equations such that the number of equations (m) is less than the number of variables (n). This causes (2) to be an *undetermined* set of linear equations which has many possible solutions. As each equation can be considered as a *hyper-*

plane in n-dimensional space, the *solution space* of the set of linear equations is, in general, a *convex polyhedron.* The computational algorithms of linear programming determine from all possible solutions one that optimizes the objective function. Since the solution space may be unbounded in the direction of optimization, the optimum value of the objective function can also be unbounded. We will only discuss the minimizing case, as maximizing a linear function is equal to minimizing the negative of the linear function.

In matrix notation the linear-programming problem is given by

minimize

$$\mathbf{cX}$$

subject to

$$\mathbf{AX} = \mathbf{b}$$
$$\mathbf{X} \geq \mathbf{0}$$

where $\mathbf{c} = (c_1,c_2, \cdots ,c_n)$ is a row vector; $\mathbf{X} = (x_1,x_2, \cdots ,x_n)$ is a column vector; $\mathbf{A}$ is the $m \times n$ matrix of coefficients; $\mathbf{b} = (a_{10},a_{20}, \cdots, a_{m0})$ is a column vector; and $\mathbf{0} = (0,0, \cdots ,0)$ is a n-rowed column vector. As it is convenient to consider the columns of $\mathbf{A}$ as points in m-dimensional space, the linear-programming problem can be written as

minimize

$$\mathbf{cX}$$

subject to

$$x_1\mathbf{P}_1 + x_2\mathbf{P}_2 + \cdots + x_n\mathbf{P}_n = \mathbf{P}_0$$
$$x_j \geq 0$$

where $\mathbf{P}_j = (a_{1j},a_{2j}, \cdots ,a_{mj})$ is a column vector for $j = 0, 1, \cdots, n$.

A *feasible solution* to the linear-programming problem is a vector $\mathbf{X} = (x_1,x_2, \cdots ,x_n)$ which satisfied conditions (2) and (3).

A *basic solution* to equations (2) is a solution obtained by setting $n - m$ variables equal to zero and solving for the remaining m variables, provided that the determinant of the coefficients of these m variables is nonzero. The m variables are called "basic variables."

A *basic feasible solution* is a basic solution which also satisfies constraints (3); i.e., all basic variables are nonnegative.

A *nondegenerate basic feasible solution* is a basic feasible solution with exactly m positive x_i.

A *minimum feasible solution* is a feasible solution which also minimizes (1).

A *basis* is a linearly independent set of vectors. A *feasible basis* for the linear-programming problem is a square matrix $\mathbf{B}$ composed of a linearly independent set of vectors selected from the rectangular matrix $\mathbf{A} = (\mathbf{P}_1\mathbf{P}_2 \cdots \mathbf{P}_n)$ such that, for the square set of equations, $\mathbf{BX}_0 = \mathbf{P}_0$, $\mathbf{X}_0 = \mathbf{B}^{-1}\mathbf{P}_0 \geq \mathbf{0}$. For example, if $\mathbf{B} = (\mathbf{P}_1\mathbf{P}_2 \cdots \mathbf{P}_m)$ we would have $\mathbf{X}_0 = (x_{10},x_{20}, \cdots x_{m0}) \geq \mathbf{0}$. Here the solution to the given problem is $\mathbf{X} = (x_{10},x_{20}, \cdots ,x_{m0},0, \cdots ,0)$, where the last $n - m$ terms of $\mathbf{X}$ are zero. Finding a feasible basis corresponds to selecting a determined square set of equations from the given underdetermined rectangular set and letting those variables not in the square set be equal to zero.

A convex combination of the vectors $\mathbf{U}_1, \mathbf{U}_2, \ldots, \mathbf{U}_n$ is a vector

$$\mathbf{U} = \alpha_1\mathbf{U}_1 + \alpha_2\mathbf{U}_2 + \cdots + \alpha_n\mathbf{U}_n$$

where the α_i are scalers, $\alpha_i \geq 0$, and $\Sigma\alpha_i = 1$.

A subset of points $\mathbf{C}$ of Euclidean space is a *convex set* if and only if, for all pairs of points $\mathbf{U}_1$ and $\mathbf{U}_2$ in $\mathbf{C}$, any convex combination

$$\mathbf{U} = \alpha_1\mathbf{U}_1 + \alpha_2\mathbf{U}_2 = \alpha\mathbf{U}_1 + (1 - \alpha)\mathbf{U}_2 \qquad 1 \geq \alpha \geq 0$$

is also in $\mathbf{C}$. A convex set is one which contains the straight line joining any two points in the set.

A point $\mathbf{U}$ in a convex set $\mathbf{C}$ is called an *extreme point* if $\mathbf{U}$ cannot be expressed as a convex combination of any other two distinct points in $\mathbf{C}$.

THEOREM 1: The set of all feasible solutions to the linear-programming problem is a convex set.

THEOREM 2: The objective function (1) assumes its minimum at an extreme point of the convex set $\mathbf{C}$ generated by the set of feasible solutions to the linear-programming problems. If it assumes its minimum at more than one extreme point, then it takes on the same value for every convex combination of those particular points.

THEOREM 3: If a set of $k \leq m$ vectors $\mathbf{P}_1, \mathbf{P}_2, \ldots, \mathbf{P}_k$ can be found that are linearly independent and such that

$$x_1\mathbf{P}_1 + x_2\mathbf{P}_2 + \cdots + x_k\mathbf{P}_k = \mathbf{P}_0$$

and all $x_i \geq 0$, then the point $\mathbf{X} = (x_1,x_2, \ldots ,x_k,0, \ldots ,0)$ is an extreme point of the convex set of feasible solutions. Here $\mathbf{X}$ is an n-dimensional vector whose last $n - k$ elements are zero.

THEOREM 4: If $\mathbf{X} = (x_1,x_2, \ldots ,x_n)$ is an extreme point of $\mathbf{C}$, then the vectors associated with positive x_i form a linearly independent set. From this it follows that, at most, m of the x_i are positive.

THEOREM 5: $\mathbf{X} = (x_1,x_2, \ldots ,x_n)$ is an extreme point of $\mathbf{C}$ if and only if the positive x_j are coefficients of linearly independent vectors $\mathbf{P}_j$ in

$$\sum_{j=1}^{n} x_j\mathbf{P}_j = \mathbf{P}_0$$

THEOREM 6: If a feasible solution exists, then a basic feasible solution exists.

THEOREM 7: If the objective function possesses a finite minimum, then at least one optimal solution is a basic feasible solution.

These theorems enable us to restrict the search for an optimum solution to extreme points of the convex set **C** of all possible solutions. A geometric discussion of the above concepts is given below in Section III.

If the linear-programming problem is stated as the following *primal problem:*

minimize

$$\mathbf{cX}$$

subject to

$$\mathbf{AX} \geq \mathbf{b}$$
$$\mathbf{X} \geq \mathbf{0},$$

then the corresponding dual problem is given by

maximize

$$\mathbf{Wb}$$

subject to

$$\mathbf{WA} \leq \mathbf{c}$$
$$\mathbf{W} \geq \mathbf{0}$$

where $\mathbf{W} = (w_1, w_2, \ldots, w_m)$ is the row vector of unknowns for the dual problem.

THE DUALITY THEOREM: If either the primal or the dual problem has a finite optimum solution, then the other problem has a finite optimum solution and the extremes of the linear functions are equal, i.e., min $\mathbf{cX} = $ max $\mathbf{Wb}$. If either problem has an unbounded optimum solution, then the other problem has no feasible solutions.

The concept of duality and the duality theorem have much significance in the theoretical and computational aspects of linear programming.

III. LINEAR-PROGRAMMING TECHNIQUES[1]

A. *The Simplex Method*

The basic computational procedure for solving *any* linear-programming problem is the *simplex method.* With the simplex method, we can, once a first basic (extreme-point) feasible solution has been determined, obtain

[1] The reader is referred to the references for further discussions of the material in Sections A and B. The remaining sections describe the material in a cursory and introductory fashion, and the reader must do additional reading in order to develop fully the material presented.

a minimum basic feasible solution in a finite number of steps. These steps, or *iterations*, consist of finding a new basic feasible solution whose corresponding value of the objective function is less than (or at worst equal to) the value of the objective function for the preceding solution. This process is continued until a minimum solution with either a finite or infinite value of the objective function has been reached. The mathematical description of the *standard simplex method* follows. The name simplex was given to this procedure as one of the first examples solved with this technique contained the inequality $x_1 + x_2 + \cdots + x_n \leq 1$, which defines a simplex (generalized tetrahedron) with unit intercepts in n-dimensional space.

Assume that all coefficients of vector $\mathbf{P}_0$ are nonnegative. We can always multiply an equation by -1 to make the corresponding $a_{i0} \geq 0$. Let $\mathbf{B}_1$ be a feasible basis; i.e., a basic feasible solution has been found from the equations $\mathbf{B}_1\mathbf{X}_{01} = \mathbf{P}_0$, where $\mathbf{X}_{01} = \mathbf{B}_1^{-1}\mathbf{P}_0 \geq 0$. In practice, $\mathbf{B}_1$ is usually a unit matrix of order m, and the corresponding first feasible solution is readily obtained since the inverse of a unit matrix is a unit matrix. For this case, $\mathbf{X}_{01} = \mathbf{P}_0$. In those situations where a suitable unit matrix is not given as part of the problem, an *artificial unit basis* is attached to the problem to start it off. This device is discussed below.

Assuming a unit basis for the first feasible solution and reordering the vectors of the basis so that $\mathbf{B}_1 = (\mathbf{P}_1\mathbf{P}_2 \cdots \mathbf{P}_m)$ (this step is not necessary, but is done here to aid in the discussion), the *simplex tableau* (computational tableau) takes the form

Basis	$\mathbf{P}_0$	$\mathbf{P}_1 \cdots \mathbf{P}_l \cdots \mathbf{P}_m$	$\mathbf{P}_{m+1} \cdots \mathbf{P}_j \cdots \mathbf{P}_k \cdots \mathbf{P}_n$
$\mathbf{P}_1$	x_{10}	$1 \cdots 0 \cdots 0$	$x_{1,m+1} \cdots x_{1j} \cdots x_{1k} \cdots x_{1n}$
$\vdots$	$\vdots$	$\vdots \quad \vdots \quad \vdots$	$\vdots \quad \vdots \quad \vdots \quad \vdots$
$\mathbf{P}_l$	x_{l0}	$0 \cdots 1 \cdots 0$	$x_{l\,m+1} \cdots x_{lj} \cdots x_{lk} \cdots x_{ln}$
$\vdots$	$\vdots$	$\vdots \quad \vdots \quad \vdots$	$\vdots \quad \vdots \quad \vdots \quad \vdots$
$\mathbf{P}_m$	x_{m0}	$0 \cdots 0 \cdots 1$	$x_{1,m+1} \cdots x_{mj} \cdots x_{mk} \cdots x_{mn}$
	x_{00}	$0 \cdots 0 \cdots 0$	$x_{0,m} \cdots x_{0j} \cdots x_{0k} \cdots x_{0n}$

where $x_{ij} = a_{ij}$, for $i = 1, \cdots m$ and $j = 0, 1, \cdots n$; the basic feasible solution is $\mathbf{X}_{01} = (x_{10}, \cdots, x_{l0}, \cdots, x_{m0}) = \mathbf{B}_1^{-1}\mathbf{P}_0$; and in general we can define $\mathbf{X}_{j1} = (x_{1j}, x_{2j} \cdots, x_{mj}) = \mathbf{B}_1^{-1}\mathbf{P}_j$. The value of the objective function is $x_{00} = \sum_{i \text{ in basis}} c_i x_{i0}$. The numbers x_{0j} for $j = 1, \cdots, n$ are defined by $x_{0j} = \sum_{i \text{ in basis}} c_i x_{ij} - c_j$. The summation term is called the indirect cost and is sometimes denoted by $z_j = \sum_{i \text{ in basis}} c_i x_{ij}$. Note that $x_{0j} = 0$

for any j in the basis. The following theorems indicate the need and the use of the x_{0j}.

THEOREM 1: If, for any j, the condition $x_{0j} > 0$ holds, then a set of feasible solutions can be constructed, such that $x'_{00} < x_{00}$ for any member of the set, where the lower bound of x'_{00} is either finite or infinite. (x'_{00} is the value of the objective function for a particular member of the set of feasible solutions.)

Case I. If the lower bound is finite, a new feasible solution consisting of exactly m positive variables can be constructed whose value of the objective function is less than the value of the preceding solution; i.e., $-\infty < x'_{00} < x_{00}$.

Case II. If the lower bound is infinite, a new feasible solution consisting of exactly $m + 1$ positive variables can be constructed whose value of the objective function can be made arbitrarily small.

THEOREM 2: If for any basic feasible solution $\mathbf{X} = (x_{10}, x_{20}, \cdots, x_{m0})$ the conditions $x_{0j} \leq 0$ hold for all $j = 1, 2, \cdots, n$, then the solution is a minimum feasible solution.

Theorem 1 assumes nondegeneracy; i.e., all basic feasible solutions to the problem are strictly positive (all $x_{i0} > 0$). The assumption is required from a theoretical point of view as it enables one to prove that the simplex method will converge in a finite number of steps. If a particular problem can have a degenerate basic feasible solution, then there is the possibility that the procedure will *cycle*, i.e., return to the same basis after a finite number of steps, and hence, not converge to the optimum solution. Although examples have been constructed which do cycle, it is quite uncommon, and a problem, degenerate or not, usually converges. Although not usually employed, computational devices exist which will guarantee the convergence of any problem.

To determine a new basic feasible solution, the following steps are carried out. These steps change the basis one vector at a time until a stop condition is encountered.

1. Compute all x_{0j}.
2. Are all $x_{0j} \leq 0$ for $j = 1, 2, \cdots, n$? (This set of inequalities is called the optimality criterion.) If yes, the current solution is an optimum solution, and the procedure stops. For any $x_{0j} = 0$ with $\mathbf{P}_j$ not in the optimum basis, an alternate optimum solution can be obtained by introducing this vector into the basis. If no, select for the vector to be introduced into the new solution the vector $\mathbf{P}_k$ whose $x_{0k} = \max_j x_{0j} > 0$. If ties occur, select any one.
3. To ensure feasibility of the new solution, the vector to be eliminated from the basis is the vector $\mathbf{P}_l$ corresponding to

$$\frac{x_{l0}}{x_{lk}} = \min_{x_{ik}>0} \frac{x_{i0}}{x_{ik}}$$

If ties occur, select any one. If all $x_{ik} \leq 0$, then the problem has an unbounded optimum solution, and the procedure stops. If the ratio $x_{l0}/x_{lk} \geq 0$ happens to equal zero (this implies the degenerate case with $x_{l0} = 0$), the value of the objective function for the new solution will be the same as before. The element x_{lk} is called the *pivot element.*

4. Determine the new solution and new simplex tableau by applying the following formulas (Gaussian elimination):

$$x'_{ij} = x_{ij} - \frac{x_{lj}x_{ik}}{x_{lk}} \qquad i \neq l$$

$$x'_{lk} = \frac{x_{ij}}{x_{lk}}$$

These formulas hold for $i = 0, 1, \cdots, m$ and $j = 0, 1, \cdots, n$. The x'_{ij} for $j = 0$ is the new basic feasible solution; x'_{00} is the new value of the objective function; x'_{0j} are the new indirect-minus-direct cost numbers. This transformation is equivalent to determining a new feasible basis $\mathbf{B}_2$ such that the new solution vector is $\mathbf{X}_{02} = \mathbf{B}_2{}^{-1}\mathbf{P}_0$ and the $\mathbf{X}_{j2} = \mathbf{B}_2{}^{-1}\mathbf{P}_j$.

The above steps are repeated for the data in the new tableau. Note that the transformation will cause the unit matrix of the initial tableau to be transformed to the inverse of the current basis.

If a unit basis is not explicitly contained in the original statement of the problem, a set of *artificial nonnegative variables* are attached to the system, one new variable for each equation. In some instances a full set of m artificial variables will not be required, as the original problem contains a partial set of unit vectors. The costs coefficients for the artificial variables are assumed to be infinite, and hence, if a minimum feasible solution to the original problem exists, the simplex method will drive the values of the artificial variables to zero. If the original problem does not have any feasible solutions, the simplex method will terminate with artificial variables in the optimum solution at a positive level. The computational tableau and steps of the simplex method can be readily modified to take care of the infinite costs.

To illustrate the above consider the following linear-programming problem, Garvin [28]. Maximize $x_1 + 2x_2$ subject to

$$\begin{array}{rrl} -x_1 + 3x_2 & \leq 10 & \quad (a) \\ x_1 + x_2 & \leq 6 & \quad (b) \\ x_1 - x_2 & \leq 2 & \quad (c) \\ x_1 + 3x_2 & \geq 6 & \quad (d) \\ 2x_1 + x_2 & \geq 4 & \quad (e) \\ x_1 & \geq 0 & \\ x_2 & \geq 0 & \end{array}$$

We convert the problem to minimization and to equalities by adding the slack variables x_3, x_4, x_5, x_6, x_7. As a full unit basis is not available, we add two artificial variables x_8 and x_9. The problem is then

$$
\begin{array}{llr}
\min = -x_1 - 2x_2 \qquad\qquad\qquad\qquad\qquad + wx_8 + wx_9 & & \\
-x_1 + 3x_2 + x_3 & = 10 & (a') \\
x_1 + x_2 \quad + x_4 & = 6 & (b') \\
x_1 - x_2 \qquad + x_5 & = 2 & (c') \\
x_1 + 3x_2 \qquad\qquad - x_6 \quad + x_8 & = 6 & (d') \\
2x_1 + x_2 \qquad\qquad\quad - x_7 \quad + x_9 & = 4 & (e') \\
x_j & \geq 0 &
\end{array}
$$

The w represents the infinite artificial cost. The first basic feasible solution is given by $x_3 = 10$, $x_4 = 6$, $x_5 = 2$, $x_8 = 6$, $x_9 = 4$; the value of the objective function is $10w$. As the artificial part of objective function and the indirect cost can be separated from the real part of these numbers, an additional row is added to the tableau, as shown below. The following five tableaus represent the complete solution to the above problem. Note that as long as artificial variables are in the solution, the vector to be introduced into the basis corresponds to the vector with the maximum artificial part of the indirect cost. As artificial variables would never be allowed to reenter a basis, they are dropped from the tableau once they are eliminated. The elements in the circles are the pivot elements. From the last (fourth) iteration, we have that the optimum solution is $x_1 = 2$; $x_2 = 4$; $x_5 = 4$; $x_6 = 8$; $x_7 = 4$; all other variables are equal to zero, and the maximum value of the objective function is $+10$.

By plotting the original inequalities in two-dimensional space as in Figure A-1, a geometric picture of the simplex method can be given.

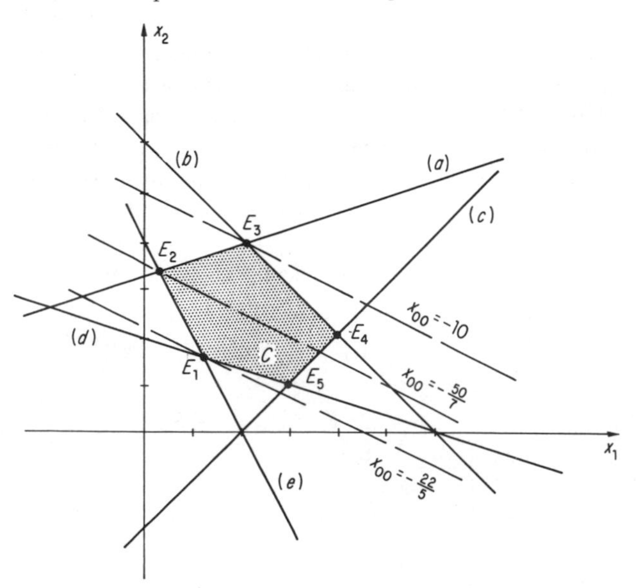

Figure A-1

	Basis	$\mathbf{P}_0$	$\mathbf{P}_1$	$\mathbf{P}_2$	$\mathbf{P}_3$	$\mathbf{P}_4$	$\mathbf{P}_5$	$\mathbf{P}_6$	$\mathbf{P}_7$	$\mathbf{P}_8$	$\mathbf{P}_9$
0	$\mathbf{P}_3$	10	-1	3	1						
	$\mathbf{P}_4$	6	1	1		1					
	$\mathbf{P}_5$	2	1	-1			1				
	$\mathbf{P}_8$	6	1	3				-1		1	
	$\mathbf{P}_9$	4	2	1					-1		1
		0	1	2							
	w	10	3	4				-1	-1		
1	$\mathbf{P}_3$	4	-2		1			1			
	$\mathbf{P}_4$	4	$2/3$			1		$1/3$			
	$\mathbf{P}_5$	4	$4/3$				1	$-1/3$			
	$\mathbf{P}_2$	2	$1/3$	1				$-1/3$			
	$\mathbf{P}_9$	2	$5/3$					$1/3$	-1		1
		-4	$1/3$					$2/3$			
	w	2	$5/3$					$1/3$	1		
2	$\mathbf{P}_3$	$32/5$			1			$7/5$	$-6/5$		
	$\mathbf{P}_4$	$16/5$				1		$1/5$	$2/5$		
	$\mathbf{P}_5$	$12/5$					1	$-3/5$	$4/5$		
	$\mathbf{P}_2$	$8/5$		1				$-2/5$	$1/5$		
	$\mathbf{P}_1$	$6/5$	1					$1/5$	$-3/5$		
		$-22/5$						$3/5$	$1/5$		
	w	0									
3	$\mathbf{P}_6$	$32/7$			$5/7$			1	$-6/7$		
	$\mathbf{P}_4$	$16/7$			$-1/7$	1			$4/7$		
	$\mathbf{P}_5$	$36/7$			$3/7$		1		$2/7$		
	$\mathbf{P}_2$	$24/7$		1	$2/7$				$-1/7$		
	$\mathbf{P}_1$	$2/7$	1		$-1/7$				$-3/7$		
		$-50/7$			$-3/7$				$5/7$		
4	$\mathbf{P}_6$	8			$1/2$	$3/2$		1			
	$\mathbf{P}_7$	4			$-1/4$	$7/4$			1		
	$\mathbf{P}_5$	4			$1/2$	$-1/2$	1				
	$\mathbf{P}_2$	4		1	$1/4$	$1/4$					
	$\mathbf{P}_1$	2	1		$-1/4$	$3/4$					
		-10			$-1/4$	$-5/4$					

The shaded area **C** is the convex set of feasible solutions, and the points $\mathbf{E}_i$ are extreme points. For this problem, the simplex method takes two steps to determine the first basic feasible solution corresponding to $\mathbf{E}_1$, moves next to $\mathbf{E}_2$, and finally to the optimum solution $\mathbf{E}_3$. The trace of the objective function through each of the extreme-point solutions is shown in the figure. In a sense, the simplex method moves the trace of the objective function from one extreme point to an adjacent extreme point until the optimum is reached. Note that if the trace of the objective function were parallel to the line joining $\mathbf{E}_3$ and $\mathbf{E}_4$, multiple optimum solutions would exist.

B. *Revised Simplex Method*

The kernel of the computational aspects of linear programming resides in the calculation of the inverse of the current basis. Given the inverse, all quantities required in a simplex iteration can be determined. Letting $\mathbf{B}_p$ be the feasible basis of the pth iteration we have

$$\begin{aligned}
\mathbf{X}_{0p} &= \mathbf{B}_p^{-1}\mathbf{P}_0 \\
\mathbf{X}_{jp} &= \mathbf{B}_p^{-1}\mathbf{P}_j \\
\boldsymbol{\pi}_p &= \mathbf{c}_p\mathbf{B}_p^{-1} \\
\boldsymbol{\pi}_p\mathbf{P}_j &= \mathbf{c}_p\mathbf{B}_p^{-1}\mathbf{P}_j \\
\boldsymbol{\pi}_p\mathbf{P}_0 &= \mathbf{c}_p\mathbf{X}_{0p} = \mathbf{c}_p\mathbf{B}_p^{-1}\mathbf{P}_0,
\end{aligned}$$

where $\mathbf{X}_{0p}$ is the solution vector for the pth basic feasible solution; $\mathbf{X}_{jp}$ are the vectors which express the given vectors $\mathbf{P}_j$ as linear combinations of the basis vectors; $\mathbf{c}_p$ is the row vector of cost coefficients of the vectors in the pth basis; the elements of the row vector $\boldsymbol{\pi}_p$ are called the "simplex multipliers"; $\boldsymbol{\pi}_p\mathbf{P}_j$ is the indirect cost of the $\mathbf{P}_j$ vector; $\boldsymbol{\pi}_p\mathbf{P}_0$ is the value of the objective function for the pth basis. From a computational point of view, using the explicit representation of the inverse and the simplex multipliers has a number of advantages. These include the reduction of the amount of computation and the reduction of the amount of information that has to be recorded per iteration. Whereas the standard simplex method transforms and records the complete simplex tableau, the revised procedure needs only to record the new inverse and solution vector. Note that the revised procedure uses the original data in each step and that if, as is the case in many problems, the data contain many zeros, computation time can be saved.

The *product-form* of the revised simplex method uses the fact that the inverse of the feasible basis, which starts out as an identity matrix, can be expressed as the product of elementary transformation matrices. Each such matrix, which differs from a unit matrix in the lth column (l corresponds to the row position of the vector eliminated from the basis) contains the information necessary to determine the row inverse. For the pth iteration let

$$\mathbf{E}_p^l = \begin{pmatrix} 1 \ldots y_{1l} \ldots 0 \\ \vdots \quad \vdots \quad \vdots \\ 0 \ldots y_{ll} \ldots 0 \\ \vdots \quad \vdots \quad \vdots \\ 0 \ldots y_{ml} \ldots 1 \end{pmatrix}$$

where

$$y_{il} = \frac{x_{ik}}{x_{lk}} \qquad i \neq l$$

$$y_{ll} = \frac{1}{x_{lk}}$$

The inverse for the pth basis is given by

$$\mathbf{E}_p^l \mathbf{E}_{p-1}^l \cdots \mathbf{E}_2^l \mathbf{E}_1 = \mathbf{B}_p^{-1}$$

with $\mathbf{E}_1 = \mathbf{I}$. With this condensed form only a limited amount of information needs to be recorded. It has been demonstrated that for most linear-programming problems, the product form of the revised simplex method is the most efficient. Computational experience has shown that the number of iterations required to find an optimum solution can vary between m and $3m$. The number appears to be more a function of the number of equations than of the number of variables. The number varies with the algorithm used, the method of finding the first feasible solution, and the criterion used to select a vector to go into the new basis.

C. *The Transportation-problem Algorithm*

Because of the structure of the matrix of coefficients which defines the transportation problem, the calculations of the simplex method as applied to the transportation problem are greatly simplified. Letting $m =$ the number of origins and $n =$ the number of destinations, each basic feasible solution of the transportation problem corresponds to an $(n + m - 1)$ triangular basis. This enables one to compute readily the solution, and since each coefficient of the basis is 0 or 1, the solution will be in integers if the original availabilities and requirements were integers. The simplex algorithms for the transportation problem utilize either the solution to the dual problem or the *stepping-stone method.* The main difference between the two techniques is the method of computing the indirect costs. The dual method is the one used in most computer codes. Other procedures exist for solving this problem, e.g., Hungarian method, flow method. A number of variations to the transportation problem exist which also have specialized computational procedures. These include the *capacitated transportation problem,* in which the amount shipped between any origin and destination has an upper bound, and the generalized transportation problem, or *machine loading problem,* in which the coefficients in the equations are not restricted to be either 0 or 1.

D. *Additional Computational Procedures*

1. THE DUAL SIMPLEX ALGORITHM

In some instances, it is easier to find a basis for which the optimality criterion is satisfied, i.e., all $x_{0j} \leq 0$, and for which the feasibility criterion is not, i.e., not all $x_{i0} \geq 0$. If this is the case, the *dual simplex algorithm* can be employed. The only difference is in the criteria used to select the vector (variable) to be introduced into the basis and the one that leaves the basis. Here the vector to leave is determined first and corresponds to

$$x_{l0} = \min x_{i0} < 0$$

The vector introduced corresponds to the index k for which

$$\frac{x_{0k} - c_k}{x_{lk}} = \min_{x_{lj}<0} \frac{x_{0j} - c_j}{x_{lj}} > 0$$

The pivot element is x_{lk}, and the elimination transformation is the same as in the standard simplex method. The above analysis causes at least one of the negative solution variables to become positive and also keeps the transformed $x_{0j} \leq 0$; i.e., the transformation improves the feasibility of the solution while still satisfying the optimality criterion.

The original simplex method (primal method) and the dual method have been combined in some instances in a *primal-dual,* or *composite simplex,* computer code. These techniques place no restrictions on the signs of the c_j or b_i and enable one to obtain an initial basic solution which either satisfies the feasibility or optimality conditions.

2. INTEGER PROGRAMMING

Many linear-programming problems require the solution to be in terms of integers, e.g., whole units to be manufactured. Unlike the transportation problem the simplex algorithm will not guarantee an integer solution for the general linear-programming problem. However, variations of the simplex method exist which will guarantee the finding of an integer optimum solution, if one exists. This integer-programming procedure adds, in a systematic fashion, new constraints, or *cutting planes,* to the original set of constraints. The new constraints change the convex set of solutions so that it will contain as an optimum extreme point a point with integer coordinates. Integer-programming codes do exist, but as the number of iterations required to solve a particular problem appears to rest heavily on the structure and data of the problem, the utilization of these codes in an operational environment is still open to question.

3. UPPER-BOUND CONDITIONS

In many instances, the variables of a linear programming are bounded from above, i.e., $x_j \leq u_j$. Again, a slight variation of the basic simplex method can solve the bounded problem without the explicit representation of the upper-bound constraints in the simplex tableau. This procedure can be used when all or some of the variables are bounded. The situation when the x_j are bounded below is easily handled by direct substitution, i.e., if $d_j \leq x_j$. Let $x_j = d_j + x_j'$ and substitute $d_j + x_j'$ for the corresponding x_j.

4. SENSITIVITY ANALYSIS

Sensitivity analysis in linear programming deals with the investigation of how the optimum solution varies with changes in the original data. For example, we are interested in how much a particular cost coefficient can vary before the computed optimum is no longer optimal, or how much a right-hand-side coefficient b_i can vary before the solution is no longer feasible, or finally, what is the effect on the optimum solution if a change is made in an a_{ij}. Techniques are available for performing these analyses and are usually included in linear-programming codes. Additional procedures exist for handling the parametric programming cases where each cost coefficient and/or each right-hand-side coefficient is a linear function of the same parameter, e.g., $c_j = d_j + \lambda d_j'$. This procedure yields sets of optimum solutions corresponding to ranges of the parameter.

5. DECOMPOSITION ALGORITHM

Although it is theoretically possible to solve any given linear-programming model, the analyst is quickly made aware of certain limitations which restrict his endeavors. Chief among these limitations is the problem of dimensionality. Almost all difficulties that arise in the development of a programming problem can be related to its size. This is certainly true for such restrictive items as the cost of data gathering, the matrix preparation, the computing of costs, the reasonableness of the linear model, etc.

For many problems the constraints consist of rather large subsets of equations which are related in that they refer to the same time period or same production facility, and these subsets are tied together by a small set of equations. These "tie-in equations" might represent total demand for a product. In problems of this sort we have, in a sense, a number of separate linear-programming problems whose joint solution must satisfy a set of additional restrictions. If we boxed in the sets of constraints and corresponding part of the objective function, Figure A-2 would result. Here we have partitioned the original problem of $\mathbf{AX} = \mathbf{b}$, $\mathbf{X} \geq \mathbf{0}$, $\mathbf{cX}$ a minimum, into the *decomposed program* of finding the vectors $\mathbf{X}_p \geq \mathbf{0}$ $(p = 0, 1, \ldots, k)$ such that

$$\sum_p \mathbf{A}_p\mathbf{X}_p = \mathbf{b}_0 \qquad (8)$$

$$\mathbf{B}_p\mathbf{X}_p = \mathbf{b}_p \qquad (9)$$

$$\sum_p \mathbf{C}_p\mathbf{X}_p \text{ is a minimum} \qquad (10)$$

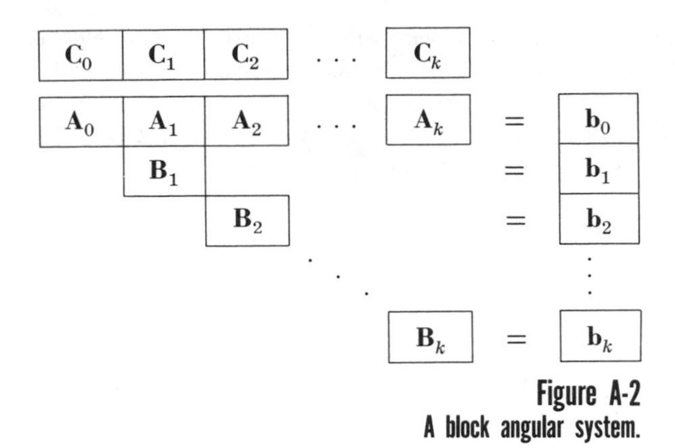

Figure A-2
A block angular system.

where $\mathbf{A}_p$ is an $m_0 \times n_p$ matrix; $\mathbf{B}_p$ is an $m_p \times n_p$ matrix; $\mathbf{C}_p$ is an n_p-vector; $\mathbf{b}_0$ is an m_0-vector; and $\mathbf{b}_p$ is an m_p-vector; $\mathbf{X}_p$ is a variable n_p-vector. As written above, this problem has $\sum_p m_p$ constraints and $\sum_p n_p$ variables. For this discussion we assume the above problem has a finite optimum.

Assume that for each p, there is available the corresponding convex set $\mathbf{S}_p$ of solutions to each subproblem $\mathbf{B}_p\mathbf{X}_p = \mathbf{b}_p$, $\mathbf{X}_p \geq \mathbf{0}$. Then the solution of the original problem could be thought of as the selection of a convex combination of solution points from $\mathbf{S}_p$ for each p so as to satisfy the tie-in restrictions $\sum_p \mathbf{A}_p\mathbf{X}_p = \mathbf{b}_0$ and make $\sum_p \mathbf{C}_p\mathbf{X}_p$ a minimum. Although this bare idea of the decomposition appears fraught with its own difficulties, there are a few saving features which enable us to consider the optimization of an $m_0 + k$ constraint problem subject to the solution of k $(m_p \times n_p)$ subproblems, instead of one large problem with $\sum_p m_p$ constraints.

The new problem to be considered is called the extremal program or master problem, and it arises in the following fashion. Let us consider for some p a particular extreme-point solution $\mathbf{X}_{pj}$ of the convex set of solutions $\mathbf{S}_p$ to the subproblem $\mathbf{X}_p \geq \mathbf{0}$, $\mathbf{B}_p\mathbf{X}_p = \mathbf{b}_p$. Define for each such extreme point j

$$\mathbf{P}_{pj} = \mathbf{A}_p\mathbf{X}_{pj} \qquad f_{pj} = \mathbf{C}_p\mathbf{X}_{pj}$$

The extremal program is to find the numbers $\lambda_{pj} \geq 0$ satisfying for all (p,j)

$$\mathbf{A}_0\mathbf{X}_0 + \sum_p\sum_j \mathbf{P}_{pj}\lambda_{pj} = \mathbf{b}_0 \tag{11}$$

$$\sum_j \lambda_{pj} = 1 \qquad (p = 1, 2, \ldots, k) \tag{12}$$

with

$$\mathbf{C}_0\mathbf{X}_0 + \sum_p\sum_j f_{pj}\lambda_{pj} \text{ a minimum} \tag{13}$$

As Dantzig and Wolfe [16] note, "The relation of the extremal problem to the original problem lies in the fact that any point of $\mathbf{S}_p$, because it is bounded (assumed) and a convex polyhedral set, may be written as a convex combination of its extreme points, that is, as $\sum_j \mathbf{X}_{pj}\lambda_{pj}$, where the λ_{pj} satisfy (12); and the expressions (11) and (13) are just the expressions (8) and (10) of the decomposed problem rewritten in terms of the λ_{pj}."

As mentioned above, this transformation yields a problem with only $m_0 + k$ constraints, the m_0 tie-in constraints (11), and the set of k single constraints (12). However, the number of variables has been increased to the total of all extreme points to the convex polyhedral $\mathbf{S}_p$, an extremely large number. The saving element of the decomposition principle is that we need only consider a small number of this total and that we only need the explicit representation of those to be considered, and only when required.

IV. SUMMARY OF LINEAR-PROGRAMMING APPLICATIONS

This section reviews the material in Chapters 1–4 and summarizes in a more formal fashion some of the important applications of linear programming, and where appropriate, a typical mathematical model is developed.

It should be stressed that the models described below, even though important in their own right, should be looked upon as starting positions or basic models in one's approach to problem formulation. The reader's problems might resemble a particular model and with suitable modifications to the basic model, a good first approximation to the real-life situation can be developed. A good example would be a transportation model which must be combined with a production-and-storage model in order to reflect the true environment. One must guard against forcing a problem to fit a particular model and, at the same time, recognize that a basic model might lend itself to the evolution of an acceptable model.

A. *Transportation Problems*

1. THE BASIC TRANSPORTATION PROBLEM

A homogeneous product is to be shipped in the amounts $a_1, a_2, \ldots, a_m$, respectively, from each of m shipping origins and received in amounts $b_1, b_2, \ldots, b_n$, respectively, by each of n shipping destinations. The cost of shipping a unit amount from the ith origin to the jth destination is c_{ij} and is known for all combinations (i, j). The problem is to determine the amounts x_{ij} to be shipped over all routes (i, j) to minimize the total cost of transportation. The a_i are the availabilities, and the b_j are the requirements.

To ensure the consistency of the mathematical model, we restrict the $\sum_i a_i = \sum_j b_j$; i.e., the sum of the availabilities equals the sum of the requirements. This is not a restriction which must be adhered to for a particular application in that if $\sum_i a_i > \sum_j b_j$, a dummy destination (e.g., a warehouse facility) is added to the problem with a requirement of $\sum_i a_i - \sum_j b_j$. If $\sum_i a_i < \sum_j b_j$ a dummy origin (e.g., purchases from a competitor) is added to the problem with an availability of $\sum_j b_j - \sum_i a_i$. Letting $x_{ij} \geq 0$ be the unknown amount to be shipped from origin i to destination j (i.e., the set of variables) the linear-programming model for $m = 2$ and $n = 3$ is then

minimize

$$c_{11}x_{11} + c_{12}x_{12} + c_{13}x_{13} + c_{21}x_{21} + c_{22}x_{22} + c_{23}x_{23}$$

subject to

$$\begin{array}{llllllll}
x_{11} + x_{12} + x_{13} & & & & & & = a_1 \\
& & & x_{21} + x_{22} + x_{23} & & & = a_2 \\
x_{11} & & + x_{21} & & & & = b_1 \\
& x_{12} & & + x_{22} & & & = b_2 \\
& & x_{13} & & + x_{23} & & = b_3 \\
& & & & & x_{ij} & \geq 0
\end{array}$$

As any one of the above equations can be deduced from the remaining $(m + n - 1)$, the rank of the system is $(m + n - 1)$. Any feasible basis is of order $(m + n - 1)$ and the corresponding matrix is triangular. This model has wide application with the sources and destinations taking on a variety of interpretations, e.g., warehouse, stores, ports, and the cost coefficients representing distances, times, dollars, etc. The model can be modified, for example, to supply information which would enable management to select from a set of possible new origins the one which is the best with respect to transportation costs. Here a set of problems would be solved, one for each possibility, with data reflecting the anticipated costs between the new origin and old destinations and the anticipated availability at the new origin and requirements for the destinations. A similar analysis would be done in order to aid in the selection of a new destination, or in the selection of an origin or destination to be closed down.

2. TRANSSHIPMENT PROBLEM

This is a transportation problem in which the origins and destinations can act as an intermediate shipping point from which the goods are transshipped to their final destination. This problem can be transformed into a basic transportation problem and can be solved by the procedures applicable to this problem.

3. CAPACITATED TRANSPORTATION PROBLEM

This is a basic transportation problem in which the possible shipments are bounded from above i.e., $x_{ij} \leq u_{ij}$. This problem can be solved with slight modification to the computational procedure for the basic problem. A special primal-dual algorithm is also available.

4. GENERALIZED TRANSPORTATION PROBLEM AND MACHINE-ASSIGNMENT PROBLEM

In this problem, the constraint set is of the form

minimize

$$\sum_i \sum_j c_{ij} x_{ij}$$

subject to

$$\sum_{j=1}^{n} a_{ij} x_{ij} = a_i \qquad (i = 1, \ldots, m) \tag{14}$$

$$\sum_{i=1}^{m} b_{ij} x_{ij} = b_j \qquad (j = 1, \ldots, n) \tag{15}$$

$$x_{ij} \geq 0$$

Problems which fit this model can be found in the transportation field and in problems of machine assignment. For the latter, the $b_{ij} = 1$, equations (14) become inequalities ($\leq$); the a_{ij} represent the time it takes to process one unit of product j on machine i; x_{ij}, the number of units of j produced on machine i; a_i, the time available on machine i; b_j, the number of units j which must be completed; and c_{ij}, the cost of processing one unit of product j on machine i.

5. MULTIDIMENSIONAL TRANSPORTATION PROBLEM

The model of this problem can take on two forms:

a. Minimize

$$\sum_i \sum_j \sum_k c_{ijk} x_{ijk}$$

subject to

$$\sum_i x_{ijk} = a_{jk}$$

$$\sum_j x_{ijk} = b_{ik}$$

$$\sum_k x_{ijk} = c_{ij}$$

$$x_{ijk} \geq 0$$

b. Minimize

$$\sum_i \sum_j \sum_k c_{ijk} x_{ijk}$$

subject to

$$\sum_j \sum_k x_{ijk} = a_i$$

$$\sum_i \sum_k x_{ijk} = b_j$$

$$\sum_i \sum_j x_{ijk} = d_k$$

$$x_{ijk} \geq 0$$

B. *Allocation or Assignment Problems*

In this problem, we have a number of individuals, machines, etc., to be assigned to perform a set of jobs. Each individual i has a given rating c_{ij} which measures his effectiveness in doing job j. An individual can be assigned to only one job. If x_{ij} represents the assignment of the ith individual to the jth job, the linear-programming formulation is then

maximize

$$\sum_i \sum_j c_{ij} x_{ij}$$

subject to

$$\sum_j x_{ij} = a_i \qquad i = 1, \ldots, m$$

$$\sum_i x_{ij} = b_j \qquad j = 1, \ldots, n$$

$$x_{ij} \geq 0$$

where a_i is the number of persons of type i available and b_j is the number of jobs of type j available. We assume $\Sigma a_i = \Sigma b_j$. In many instances all a_i and b_j equal 1 and $m = n$. Mathematically, the assignment problem is a transportation problem.

C. *Tanker-routing Problem*

Given a number of loading points $i = 1, 2, \ldots, m$ at which tankers are to be loaded for delivery to destination points $j = 1, 2, \ldots, n$. We know the time at which a tanker is to be loaded at i for making a delivery to j and the time it takes for a tanker to go from any point to any other. (We assume all tankers are identical and delivery is in units of tanker capacity.) Having made a delivery a tanker can travel to any loading point. The problem is to devise routes for each tanker so that the given loading schedule is met with the minimum number of tankers.

D. *Network-flow Problems*

Given a network (road, railway, pipeline, etc.) consisting of a single source (origin), a single sink (destination), and intermediate nodes (transfer points), let x_{ij} be the flow between any point i to point j of the network. A point is either the source, sink, or any node. We assume that the network is oriented and that there is a capacity restriction $f_{ij} \geq 0$ on each branch (i, j). We wish to determine the flow f through the network which maximizes the flow from source to sink.

For a general network, with $i = 0$ the source node and $i = m$ the sink node, we have the following formulation:

maximize

$$f$$

subject to

$$\sum_j (x_{oj} - x_{jo}) = f \tag{16}$$

$$\sum_j (x_{ij} - x_{ji}) = 0 \qquad i \neq 0, m \tag{17}$$

$$\sum_j (x_{mj} - x_{jm}) = -f \tag{18}$$

$$0 \leq x_{ij} \leq f_{ij} \tag{19}$$

(17) represents the conservation of flow at the nodes; i.e., what flows in must flow out and the summations are taken over those j which form branches for the given i. The objective function maximizes the flow out of the source or the flow into the sink, (16) and (18).

E. *Bid Evaluation*

Whenever a government agency wishes to procure items from civilian sources, producers of the items must be invited to participate in the bidding of contracts. The individual manufacturer submits bids which reflect his desire for profit, his guess about the other fellow's bid, and his own peculiar limitations. The agency must award contracts in such a way that the total dollar cost to the government is at a minimum. These types of problems can be transformed into a sequence of transportation problems and solved by that algorithm.

F. *Activity-analysis Problems*

A manufacturer has at his disposal fixed amounts of a number of different resources. These resources, such as raw material, labor, and equipment, can be combined to produce any one of several different commodities or combinations of commodities. The manufacturer knows how much of resource i it takes to produce one unit of commodity j. He also knows how much profit he makes for each unit of commodity j produced. The manufacturer desires to produce that combination of commodities which will maximize the total profit. For this problem, we define the following:

$m =$ the number of resources
$n =$ the number of commodities
$a_{ij} =$ the number of units of resource i required to produce one unit of the commodity j
$b_i =$ the maximum number of units of resource i available
$c_j =$ profit per unit of commodity j produced
$x_j =$ the level of activity (the amount produced) of the jth commodity

The a_{ij} are sometimes called input-output coefficients.

The total amount of the ith resource that is used is given by the linear expression

$$a_{i1}x_1 + a_{i2}x_2 + \dots + a_{in}x_n$$

Since this total amount must be less than or equal to the maximum number of units of the ith resource available, then we have, for each i, a linear inequality of the form

$$a_{i1}x_1 + a_{i2}x_2 + \cdots + a_{in}x_n \leq b_i$$

Since a negative x_j has no appropriate interpretation, we require that all $x_j \geq 0$. The profit derived from producing x_j units of the jth commodity is given by c_jx_j. This formulation yields the problem of maximizing the profit function

$$c_1x_1 + c_2x_2 + \cdots + c_nx_n$$

subject to the conditions

$$\begin{array}{ccccccc} a_{11}x_1 & + & a_{12}x_2 & + \cdots + & a_{1n}x_n & \leq & b_1 \\ a_{21}x_1 & + & a_{22}x_2 & + \cdots + & a_{2n}x_n & \leq & b_2 \\ \vdots & & \vdots & & \vdots & & \vdots \\ a_{m1}x_1 & + & a_{m2}x_2 & + \cdots + & a_{mn}x_n & \leq & b_m \end{array}$$

and

$$\begin{array}{llll} x_1 & & & \geq 0 \\ & x_2 & & \geq 0 \\ & & \ddots & \vdots \\ & & x_n & \geq 0 \end{array}$$

These problems arise in the field of economics in discussions on the theory of the firm and in interindustry (input-output) analysis.

G. *The Diet Problem*

Here we are given the nutrient content of a number of different foods. For example, we might know how many milligrams of phosphorus or iron are contained in one ounce of each food being considered. We are also given the minimum daily requirement for each nutrient. Since we know the cost per ounce of food, the problem is to determine the diet that satisfies the minimum daily requirements and is also the minimum cost diet. Define

m = the number of nutrients
n = the number of foods
a_{ij} = the number of milligrams of the ith nutrient in one ounce of the jth food
b_i = the minimum number of milligrams of the ith nutrient required
c_j = the cost per ounce of the jth food
x_j = the number of ounces of the jth food to be purchased ($x_j \geq 0$)

The total amount of the ith nutrient contained in all the purchased food is given by

$$a_{i1}x_1 + a_{i2}x_2 + \cdots + a_{in}x_n$$

Since this total amount must be greater than or equal to the minimum daily requirement of the ith nutrient, this linear-programming problem involves minimizing the cost function

$$c_1x_1 + c_2x_2 + \cdots + c_nx_n$$

subject to the conditions

$$\begin{array}{l} a_{11}x_1 + a_{12}x_2 + \cdots + a_{1n}x_n \geq b_1 \\ a_{21}x_1 + a_{22}x_2 + \cdots + a_{2n}x_n \geq b_2 \\ \vdots \\ a_{m1}x_1 + a_{m2}x_2 + \cdots + a_{mn}x_n \geq b_m \end{array}$$

and

$$\begin{array}{l} x_1 \geq 0 \\ x_2 \geq 0 \\ \vdots \\ x_n \geq 0 \end{array}$$

Although questionable when applied to diets for human consumption, this formulation has proved highly satisfactory for the evaluation of diets for cattle and chickens.

H. *Blending Problem*

Blending problems refer to situations where a number of components are mixed together to yield one or more products. There are restrictions on the available quantities of raw materials, restrictions on the quality of the products, and restrictions on the quantities of the products to be produced. There are usually infinitely different ways in which the raw materials can be blended to form the final products while satisfying the various constraints. It is desired to carry out the blending operation so that the given objective function can be optimized. Blending problems are found in the making of petroleum (blending of gasolines), paint, steel, etc. A description of the mathematical model would call for too specific a discussion of a particular process and, hence, is beyond the scope of this presentation. We refer the interested reader to the bibliography.

I. *Production Scheduling*

A manufacturer knows that he has to produce r_i $(i = 1, \ldots, n)$ items of a certain commodity during the next n months. They can be produced either in regular time, subject to a ceiling of a_i per month, or in overtime, subject to a ceiling of b_i per month. The cost of one item produced in the ith month is c_i on regular time and d_i on overtime. The variation of cost with time and also the capacity restrictions might make it more economical to produce in advance of the period when the items are actually needed. Storage cost is assumed to be s_i per item in each month. We wish to determine the production schedule which minimizes the sum of production and storage costs. Although this problem can be formulated as a standard linear-programming model, it can also be formulated in terms of a transportation problem and solved by that algorithm.

J. *Smooth Patterns of Production*

A manufacturer of an item must determine his monthly production schedule for the next n months. The demand for his product fluctuates, but he must always meet monthly requirements as given to him by his sales forecast or advance orders. He can fulfill the individual demands either by producing the desired amount during the month or by producing part of the desired amount and making up the difference by using the overproduction from previous months.

In general, any such scheduling problem has many different schedules that will satisfy the requirements. For example, the manufacturer could produce each month the exact number of units required by the sales forecast. However, since a fluctuating production schedule is costly to maintain, this type of production schedule is not an efficient one. On the other hand, the manufacturer with fluctuating requirements could overproduce in periods of low requirements, store the surplus, and use the excess in periods of high requirements. The production pattern can thus be made quite stable. However, because of the cost of keeping a manufactured item in storage, such a solution may be undesirable if it yields comparatively large monthly surpluses. For this problem, we wish to determine a production schedule that minimizes the sum of the costs due to output fluctuations and to inventories.

Letting

x_t = the production in month t
r_t = the requirements in month t
s_t = the storage in month t

The linear-programming formulation becomes

$$a\sum_t y_t + b\sum_t s_t$$

subject to

$$x_t + s_{t-1} - s_t = r_t \tag{20}$$
$$x_t - x_{t-1} - y_t + z_t = 0 \tag{21}$$

where $(x_t, s_t, y_t, z_t) \geq 0$; (20) states that the amount produced in month t plus the previous month's storage equals the requirements and the storage in month t; (21) states that the difference between the production in month t and the production in month $(t - 1)$ can be represented as the difference between two nonnegative numbers with y_t representing an increase in production and z_t a decrease; a is the cost of a unit of increase in production, and b is the cost of storing a unit for one month.

K. *Trim-loss Problem*

Paper mills produce rolls of a given, standard width. Customers require rolls of various widths, and hence, the rolls of standard width must be cut. In general, some waste occurs at the end of the cutting process, i.e., trim loss. The manufacturer wishes to cut the rolls as ordered by his customers and to minimize the total trim loss. This application applies to similar manufacturing situations in which a standard roll, sheet, etc., must be cut with resulting trim loss.

L. *Caterer's Problem*

A caterer knows that, in connection with the meals he has arranged to serve during the next n days, he will need r_j fresh napkins on the jth day, with $j = 1, 2, \ldots, n$. Laundering normally takes p days; i.e., a soiled napkin sent for laundering immediately after use on the jth day is returned in time to be used again on the $(j + p)$th day. However, the laundry also has a higher-cost service which returns the napkins in $q < p$ days (p and q integers). Having no usable napkins on hand or in the laundry, the caterer will meet his early needs by purchasing napkins at a cents each. Laundering expense is b and c cents per napkin for the normal and high-cost service, respectively. How does he arrange matters to meet his needs and minimize his outlays for the n days?

M. *Traveling-salesman Problem*

The problem is to find the shortest route for a salesman starting from a given city, visiting each of the specified group of cities, and then returning to the original point of departure. The linear-programming formulation of this problem requires the variables to be integers.

N. *Warehouse Problem*

Given a warehouse with fixed capacity and initial stock of a certain product, which is subject to known seasonal price and cost variations, and given a delay between the purchasing and the receiving of the product, what is the pattern of purchasing, storage, and sales which maximizes profit over a given period of time?

O. *Critical-path Planning and Scheduling*

A characteristic of many projects is that all work must be performed in some well-defined order; e.g., in construction work forms must be built before concrete can be poured. This formulation concerns the scheduling of the jobs which combine to make a project. The analysis requires a graphical representation of a project for which the cost and starting and ending times for each job of the project are known. The linear-programming formulation provides a means of selecting the least costly schedule for desired and feasible project-completion time.

P. *Structural-design Problems*

A class of problems in structural design, e.g., the automatic plastic design of structural frames, can be treated by linear programming. As a discussion of these problems is too specialized, the interested reader is referred to the bibliography.

Q. *Zero-sum Two-person Games*

It should be noted that all zero-sum two-person games can be represented as a linear-programming problem and that a linear program can also be transformed into a zero-sum two-person game. For a game, the dual of the linear-programming formulation for one player represents the linear-programming formulation for the second player. The computational procedures from one area can be applied to solve a problem from the other area.

REFERENCES

1. Ackoff, R. L.: The Development of Operations Research as a Science, *Journal of the Operations Research Society of America,* vol. 4, no. 3, June, 1956.
2. Allen, R. G. D.: "Mathematical Economics," The Macmillan Company, New York, 1956.
3. Bellmore, M., and G. L. Nemhauser: The Traveling Salesman Problem: A Survey, *Operations Research,* vol. 16, no. 3, June, 1968.
4. Charnes, A., and W. W. Cooper: "Management Models and Industrial Application of Linear Programming," John Wiley & Sons, Inc., New York, 1960.
5. Charnes, A., and W. W. Cooper: The Stepping Stone Method of Explaining Linear Programming Calculations in Transportation Problems, *Management Science,* vol. 1, 1954–1955.
6. Charnes, A., W. W. Cooper, and A. Henderson: "Introduction to Linear Programming," John Wiley & Sons, Inc., New York, 1953.
7. Charnes, A., W. W. Cooper, and B. Mellon: Blending Aviation Gasolines—A Study in Programming Interdependent Activities in an Integrated Oil Company, *Econometrica,* vol. 20, 1952.
8. Churchman, C. W., R. L. Ackoff, and E. L. Arnoff: "Introduction to Operations Research," John Wiley & Sons, Inc., New York, 1957.
9. Cooper, W. W., and A. Charnes: Linear Programming, *Scientific American,* August, 1954.
10. Dantzig, G. B.: Maximization of a Linear Function of Variables Subject to Linear Inequalities, chap. 21 of Koopmans [41].
11. Dantzig, G. B.: Application of the Simplex Method to a Transportation Problem, chap. 23 of Koopmans [41].
12. Dantzig, G. B.: "Computational Algorithm of the Revised Simplex Method," *RAND Report RM-1266,* The RAND Corporation, Santa Monica, Calif., 1953.
13. Dantzig, G. B.: "On Integer and Partial Integer Linear Programming Problems," *RAND Report P-1410,* The RAND Corporation, Santa Monica, Calif., 1958.
14. Dantzig, G. B.: *"Linear Programming,"* Princeton University Press, Princeton, N.J., 1963.
15. Dantzig, G. B., W. Orchard-Hays, and G. Waters: "Product-form Tableau for Revised Simplex Method," *RAND Report RM-1268A,* The RAND Corporation, Santa Monica, Calif., 1954.
16. Dantzig, G. B., and P. A. Wolfe: "A Decomposition Principle for Linear Programs," *RAND Report P-1544,* The RAND Corporation, Santa Monica, Calif., 1959; also in *Operations Research,* vol. 8, no. 1, 1959.
17. Dennis, J. B.: A High-speed Computer Technique for the Transportation Problem, *Journal of the Association for Computing Machinery,* vol. 5, no. 2, 1958.
18. Dennis, J. B.: "Mathematical Programming and Electrical Networks," John Wiley & Sons, Inc., New York, 1959.
19. Dickson, J. C., and F. P. Frederick: A Decision Rule for Improved Efficiency in Solving Linear Programming Problems with the Simplex Algorithm, *Communications of the ACM,* vol. 3, no. 9, 1960.
20. Dorfman, R.: "Application of Linear Programming to the Theory of the Firm," University of California Press, Berkeley, 1951.
21. Dorfman, R.: Mathematical, or "Linear," Programming, *American Economic Review,* vol. 43, December, 1953.
22. Dorfman, R., P. A. Samuelson, and R. Solow: "Linear Programming and Economic Analysis," McGraw-Hill Book Company, New York, 1958.
23. Eisemann, K., and J. R. Lourie: "The Machine Loading Problem," IBM Applications Library, New York, 1959.

24. Flood, M. M.: A Transportation Algorithm and Code, *Preprint No. 44*, University of Michigan, Ann Arbor, 1960.
25. Ford, L. R., Jr., and D. R. Fulkerson: "Solving the Transportation Problem," *RAND Report RM-1736*, The RAND Corporation, Santa Monica, Calif., 1956.
26. Gale, D.: "The Theory of Linear Economic Models," McGraw-Hill Book Company, New York, 1960.
27. Gale, D., H. W. Kuhn, and A. W. Tucker: Linear Programming and the Theory of Games, chap. 19 of Koopmans [41].
28. Garvin, W. W.: "Introduction to Linear Programming," McGraw-Hill Book Company, New York, 1960.
29. Gass, S. I.: "Linear Programming: Methods and Applications," 3d ed., McGraw-Hill Book Company, 1969.
30. Gass, S. I.: Recent Developments in Linear Programming, "Advances in Computers," vol. 2, Academic Press, New York, 1961.
31. Goldman, A. J., and A. W. Tucker: Theory of Linear Programming, pp. 53–97 of Kuhn and Tucker [42].
32. Gomory, R. E.: "All-integer Programming Algorithm," *Report RC-189*, IBM Research Center, Mohansic, New York, 1960.
33. Gomory, R. E.: "An Algorithm for Integer Solutions to Linear Programs," *Princeton-IBM Mathematical Research Project Technical Report No. 1*, Princeton, N.J., 1958.
34. Gomory, R. E.: "An Algorithm for the Mixed Integer Problem," *RM-2597*, The RAND Corporation, Santa Monica, Calif., 1960.
35. Goode, H. W.: The Application of a Highspeed Computer to the Definition and Solution of the Vehicular Traffic Problem, *Journal of the Operations Research Society*, vol. 5, no. 6, Dec., 1957.
36. Hadley, G.: "Linear Programming," Addison-Wesley Publishing Company, Inc., Reading, Mass., 1962.
37. Heyman, J., and W. Prager: Automatic Minimum Weight Design of Steel Frames, *Journal of the Franklin Institute*, vol. 266, no. 5, 1958.
38. Kalker, J. J.: Automatic Minimum Weight Design of Steel Frames on the IBM 704 Computer, *Report IBM 2038/3*, Brown University, Providence, R.I., 1958.
39. Kelley, J. E., Jr.: "The Cutting-plane Method for Solving Convex Programs," Mauchly Associates, Ambler, Pa., 1959.
40. Kelley, J. E., Jr.: Parametric Programming and Primal-Dual Algorithm, *Operations Research*, vol. 7, no. 3, 1959.
41. Koopmans, T. C. (ed.): "Activity Analysis of Production and Allocation," *Cowles Commission Monograph* 13, John Wiley & Sons, Inc., New York, 1951.
42. Kuhn, H. W., and A. W. Tucker: "Linear Inequalities and Related Systems," *Annals of Mathematics Studies* 38, Princeton University Press, Princeton, N.J., 1956.
43. Lemke, C. E.: The Dual Method of Solving the Linear Programming Problem, *Naval Research Logistics Quarterly*, vol. 1, no. 1, 1954.
44. Lemke, C. E., A. Charnes, and O. C. Sienkiewicz: "Plastic Limit Analysis and Integral Linear Programs," *Mathematical Report No. 21*, Rensselaer Polytechnical Institute, Troy, N.Y., 1959.
45. Luce, R. D., and H. Raiffa: "Games and Decisions," John Wiley & Sons, Inc., New York, 1957.
46. McKinsey, J. C. C.: "Introduction to the Theory of Games," McGraw-Hill Book Company, 1952.
47. Orchard-Hays, W.: "Background, Development and Extensions of the Revised

Simplex Method," *RAND Report RM-1433,* The RAND Corporation, Santa Monica, Calif., 1954.

48. Orchard-Hays, W.: "The RAND Code for the Simplex Method," *RAND Report RM-1269,* The RAND Corporation, Santa Monica, Calif., 1954.
49. Orchard-Hays, W.: "'SCROL,' A Comprehensive Operating System for Linear Programming on the IBM 704," CEIR, Arlington, Va., 1960.
50. Riley, V., and S. I. Gass: "Bibliography on Linear Programming and Related Techniques," Johns Hopkins Press, Baltimore, 1958.
51. Saaty, T. L.: "Mathematics Methods of Operations Research," McGraw-Hill Book Company, New York, 1959.
52. Tucker, A. W.: An Integer Program for a Multiple-trip Variant of the Traveling Salesman Problem, Princeton University, Princeton, N.J., 1960; *Journal of the Association for Computing Machinery,* vol. 7, no. 4, 1960.
53. Vajda, S.: "An Introduction to Linear Programming and the Theory of Games," John Wiley & Sons, Inc., New York, 1960.
54. Vajda, S.: "Mathematical Programming," Addison-Wesley Publishing Company, Inc., Reading, Mass., 1958.
55. Vajda, S.: "Readings in Mathematical Programming," John Wiley & Sons, Inc., New York, 1962.
56. Wagner, H. M.: A Comparison of the Original and Revised Simplex Methods, *Operations Research,* vol. 5, no. 3, 1957.
57. Wagner, H. M.: A Practical Guide to the Dual Theorem, *Operations Research,* vol. 6, no. 3, 1958.
58. Wagner, H. M.: The Simplex Method for Beginners, *Operations Research,* vol. 6, no. 2, 1958.
59. Wagner, H. M.: The Dual Simplex Algorithm for Bounded Variables, *Naval Research Logistics Quarterly,* vol. 5, no. 3, 1958.
60. Wolfe, P. (ed.): *Linear Programming and Recent Extensions,* RAND Symposium on Mathematical Programming, The RAND Corporation, Santa Monica, Calif., 1959.
61. Wolfe, P.: The Composite Simplex Algorithm, *SIAM Review,* vol. 7, no. 1, 1965.

BIBLIOGRAPHY OF LINEAR-PROGRAMMING APPLICATIONS[1]

[1] From Saul I. Gass, *Linear Programming*, 3d ed., McGraw-Hill Book Company, New York, 1969.

1. AGRICULTURAL APPLICATIONS

Boles, James N.: Linear Programming and Farm Management Analysis, *Journal of Farm Economics*, vol. 37, no. 1, pp. 1–24, February, 1955.

Candler, Wilfred: A Modified Simplex Solution for Linear Programming with Variable Capital Restriction, *Journal of Farm Economics*, vol. 38, no. 4, pp. 940–955, November, 1956.

Fisher, Walter D., and Leonard W. Shruben: Linear Programming Applied to Feed-mixing under Different Price Conditions, *Journal of Farm Economics*, vol. 35, no. 4, pp. 471–483, November, 1953.

Fox, Karl A., and Richard C. Taeuber: Spatial Equilibrium Models of the Livestock-feed Economy, *The American Economic Review*, vol. 45, no. 4, pp. 584–608, September, 1955.

———: "A Spatial Equilibrium Model of the Livestock-feed Economy in the United States." Paper presented before the meeting of the Econometric Society, Dec. 27, 1952, Chicago, Ill.; published in *Econometrica*, vol. 21, no. 4, pp. 547–566 (including references), October, 1953.

Heady, E. O., and A. C. Egbert: Regional Programming of Efficient Agricultural Production Patterns, *Econometrica*, vol. 32, no. 3, July, 1964.

Hildreth, Clifford G.: "Economic Implications of Some Cotton Fertilizer Experiments." Paper presented at a joint meeting of the Econometric Society and the American Farm Economic Association, December, 1953, and at a Cowles Commission staff meeting January, 1954; published in *Econometrica*, vol. 23, no. 1, pp. 88–98, January, 1955.

———, and Stanley Reiter: On the Choice of a Crop Rotation Plan, chap. 11 (pp. 177–188) in Tjalling C. Koopmans (ed.), "Activity Analysis of Production and Allocation," *Cowles Commission Monograph* 13 (proceedings of a Conference on Linear Programming held in Chicago, Ill. by the Cowles Commission for Research in Economics, June 20–24, 1949), John Wiley & Sons, Inc., New York, 1951.

King, Richard A.: Use of Economic Models: Some Applications of Activity Analysis in Agricultural Economics, North Carolina Agricultural Experiment Station Journal Paper 508, *Journal of Farm Economics*, vol. 35, no. 5, pp. 823–833, December, 1953.

Pierson, D. R.: Farm Profits Up by 40 Percent, *Automatic Data Processing*, vols. 4, 5, May, 1962.

Swanson, Earl R.: Integrating Crop and Livestock Activities in Farm Management Activity Analysis, *Journal of Farm Economics*, vol. 37, no. 5, pp. 1249–1258, December, 1955.

Swanson, L. W., and J. G. Woodruff: A Sequential Approach to the Feed-mix Problem, *Operations Research*, vol. 12, no. 1, January–February, 1964.

2. CONTRACT AWARDS

Percus, Jerome, and Leon Quinto: The Application of Linear Programming to Competitive Bond Bidding, *Econometrica*, vol. 24, no. 4, pp. 314–428 (including references), October, 1956.

Stanley, E. D., D. Honig, and L. Gainen: Linear Programming in Bid Evaluation, *Naval Research Logistics Quarterly*, vol. 1, no. 1, 1954.

Waggener, H. A., and G. Suzuki: Bid Evaluation for Procurement of Aviation Fuel at DFSC: A Case History, *Naval Research Logistics Quarterly*, vol. 14, no. 1, March, 1967.

3. INDUSTRIAL APPLICATIONS

a. Chemical Industry

Arnoff, E. Leonard: "The Application of Linear Programming." Paper presented at the Case Institute of Technology, Cleveland, Ohio, Jan. 20–22, 1954; published in *Proceedings of Conference on Operations Research in Production and Inventory Control*, pp. 47–52, Case Institute of Technology, Cleveland, Ohio, 1954.

Dantzig, G. B., S. Johnson, and W. White: A Linear Programming Approach to the Chemical Equilibrium Problem, *Management Science*, vol. 5, no. 1, October, 1958.

b. The Coal Industry

Henderson, James M.: A Short-run Model for the Coal Industry, *Review of Economics and Statistics*, vol. 37, no. 4, pp. 336–346, November, 1955.

c. Commercial Airlines

Morton, George: "Application of Linear Programming Methods to Commercial Airline Operations." Paper presented to the fourteenth European meeting of the Econometric Society, Cambridge, England, Aug. 13–15, 1952; abstracted in *Econometrica*, vol. 21, no. 1, p. 193 (with discussion), January, 1953.

d. Communications Industry

Kalaba, Robert E., and Mario L. Juncosa: "Optical Design and Utilization of Communication Networks." Paper presented before a joint meeting of the Institute of Management Sciences and the Operations Research Society of America, University of California, Los Angeles, May 30, 1956; published as P–782, The RAND Corporation, 25 pp., July 13, 1956; also RM–1687, 22 pp., Apr. 23, 1956; published in *Management Science*, vol. 3, no. 1, pp. 33–44 (including references), October, 1956.

Saaty, T. L., and G. Suzuki: A Nonlinear Programming Model in Optimum Communication Satellite Use, *SIAM Review*, vol. 7, no. 3, July, 1965.

e. Iron and Steel Industry

Fabian, Tibor: "Application of Linear Programming to Steel Production Planning." Paper presented to the seventh national meeting of the Operations Research Society of America, Los Angeles, Aug. 15–17, 1955; abstracted in *Journal of the Operations Research Society of America*, vol. 3, no. 4, p. 565, November, 1955.

———: Blast Furnace Burdening and Production Planning, *Management Science*, vol. 14, no. 2, October, 1967.

Reinfeld, Nyles V.: Do You Want Production or Profit? *Tooling and Production*, vol. 20, no. 5, pp. 44–48, 69, August, 1954.

f. Paper Industry

Gilmore, P. C., and R. E. Gomory: A Linear-programming Approach to the Cutting-stock Problem–Part I, *Operations Research,* vol. 9, no. 6, November–December, 1961.

Gilmore, P. C., and R. E. Gomory: A Linear-programming Approach to the Cutting-stock Problem–Part II, *Operations Research,* vol. 11, no. 6. November–December, 1963.

Morgan, J. I.: Survey of Operations Research, *Paper Mill News,* vol. 84, no. 11, March, 1961.

Paull, A. E.: "Linear Programming: A Key to Optimum Newsprint Production." Paper presented to the summer meeting of the Technical Section, Canadian Pulp and Paper Association, Quebec, Canada, June 6–8, 1955; published in *Pulp and Paper Magazine of Canada,* vol. 57, no. 1, pp. 85–90, January, 1956, reissued, vol. 57, no. 4, pp. 145–150, March, 1956.

g. Petroleum Industry

Aronofsky, J. S., and A. C. Williams: The Use of Linear Programming and Mathematical Models in Underground Oil Production, *Management Science,* vol. 8, no. 4, July, 1962.

Catchpole, A. R.: The Application of Linear Programming to Integrated Supply Problems in the Oil Industry, *Operational Research Quarterly* (U.K.), vol. 13, no. 2, June, 1962.

Charnes, Abraham, William W. Cooper, and Robert Mellon: "Blending Aviation Gasolines—A Study in Programming Interdependent Activities in an Integrated Oil Company." Paper presented to the Symposium on Linear Inequalities and Programming, Washington, D.C., June 14–16, 1951, jointly sponsored by the Air Force, DCS/Comptroller, Headquarters USAF, and the National Bureau of Standards; published in Project SCOOP, Manual 10, pp. 115–146 (including references), April, 1952; also published in *Econometrica,* vol. 20, no. 2, pp. 135–159, April, 1952; abstracted in *Operational Research Quarterly,* vol. 3, no. 3, pp. 54–55, September, 1952.

Faur, P.: Elements for Selection of Investments in the Refining Industry, *Proceedings of the Second International Conference on Operational Research,* English Universities Press, Ltd., London, and John Wiley & Sons, Inc., New York, 1960.

Garvin, W. W., H. W. Crandell, J. B. John, and R. A. Spellman: Applications of Linear Programming in the Oil Industry, *Management Science,* vol. 3, no. 4, July, 1957.

Manne, Alan S.: "Concave Programming for Gasoline Blends," P-383, The RAND Corporation, Mar. 20, 1953. Paper presented to the 1953 annual meeting of the Operations Research Society of America, Case Institute of Technology, Cleveland, Ohio, May 15–16, 1953; abstracted in *Journal of the Operations Research Society of America,* vol. 1, no. 3, p. 148, May, 1953.

———: "Scheduling of Petroleum Refinery Operations," 181 pp., Harvard Economic Studies, vol. 48, Harvard University Press, Cambridge, Mass., 1956.

Symonds, Gifford H.: "A Crude Allocation Problem." A Manufacturing Technical Committee Report of Esso Standard Oil Company, Linden, N.J.; published as chap. 9 (pp. 63–74) of "Linear Programming: The Solution of Refinery Problems," Esso Standard Oil Company, New York, 1955.

———: "Linear Programming: The Solution of Refinery Problems," 74 pp., Esso Standard Oil Company, New York, 1955.

———: "Optimum Production Rates and Inventory to Meet Uncertain Seasonal Requirements." A Manufacturing Technical Committee Report of Esso Standard Oil Company, Linden, N.J., published as chap. 6 (pp. 37–46) of "Linear Programming: The Solution of Refinery Problems," Esso Standard Oil Company, New York, 1955.

Vazsonyi, Andrew: Optimizing a Function of Additively Separated Variables Subject to a Simple Restriction, *Proceedings of the Second Symposium in Linear Programming*, vol. II, pp. 453–469 (including references) (collection of papers presented to a conference sponsored by the National Bureau of Standards and the Directorate of Management Analysis Service, DCS/Comptroller, Headquarters USAF, held in Washington, D.C., Jan. 27–29, 1955).

H. Railroad Industry

Charnes, Abraham, and M. H. Miller: "A Model for Optimal Programming of Railway Freight Train Movements." Paper presented to the meeting of the Econometric Society, New York City, Dec. 28–30, 1955; abstracted in *Econometrica*, vol. 24, no. 3, pp. 349–350, July, 1956; full article published in *Management Science*, vol. 3, no. 1, pp. 74–93 (including references), October, 1956.

Crane, Roger R.: A New Tool: Operations Research, *Modern Railroads*, vol. 9, no. 1, pp. 146–152, January, 1954.

i. Other Industries

Androit, J., and J. Gaussens: Programme for Thermal Reactor and "Breeder Reactor" Power Stations—Fuel Economy Problem of Storage—Price of Plutonium, *Proceedings of the Second International Conference on Operational Research*, English Universities Press, Ltd., London, and John Wiley & Sons, Inc., New York, 1960.

Charnes, Abraham, William W. Cooper, and Robert O. Ferguson: Optimal Estimation of Executive Compensation by Linear Programming, *Management Science*, vol. 1, no. 2, pp. 138–151, January, 1955.

Horowitz, J., R. Lattes, and E. Parker: Some Operational Problems Connected with Power Reactor Discharges, *Proceedings of the Second International Conference on Operational Research*, English Universities Press, Ltd., London, and John Wiley & Sons, Inc., New York, 1960.

4. ECONOMIC ANALYSIS

Altschul, Eugen: "Reorientation in Economic Theory: Linear and Nonlinear Programming." Paper presented to the annual spring meeting of the Missouri Section of the Mathematical Association of America, University of Kansas City, Kansas City, Mo., Apr. 22, 1955; abstracted in *American Mathematical Monthly*, vol. 62, no. 6, p. 543, June, 1955.

Beckmann, Martin J., and Thomas Marschak: "An Activity Analysis Approach to Location Theory," 38 pp., P-649, The RAND Corporation, Apr. 5, 1955. Published in *Proceedings of the Second Symposium in Linear Programming*, vol. I, pp. 331–379 (collection of papers presented to a conference sponsored jointly by the National Bureau of Standards and the Directorate of Management Analysis Service, DCS/Comptroller, Headquarters USAF, held in Washington, D.C., Jan. 27–29, 1955).

Brown, J. A. C.: "An Experiment in Demand Analysis: The Computation of a Diet Problem on the Manchester Computer." Paper presented to the Conference on Linear Programming arranged by Ferranti, Ltd., held in London, May 4, 1954; published in *Conference on Linear Programming*, pp. 41–53, Ferranti, Ltd., London, 1954; summary in *Operations Research* (*JORSA*), vol. 4, no. 1, p. 133, February, 1956.

Charnes, Abraham, and William W. Cooper: "An Example of Constrained Games in Industrial Economics." Presented to a meeting of the Econometric Society, Washington, D.C., Dec. 27–29, 1953; published as *ONR Research Memorandum* 12, Graduate School of Industrial Administration, Carnegie Institute of Technology, Pittsburgh, Pa.; abstracted in *Econometrica*, vol. 22, no. 4, pp. 526–527, October, 1954.

Chipman, John: Linear Programming, *The Review of Economics and Statistics*, vol. 35, no. 2, pp. 101–117, May, 1953.

Davidson, Donald, and Patrick Suppes: "Experimental Measurement of Utility by Use of a Linear Programming Model," 30 pp., *Technical Report* 3, Applied Mathematics and Statistical Laboratory, Stanford University, Apr. 6, 1956. Paper presented to the meeting of the Econometric Society, Ann Arbor, Mich., Aug. 29–Sept. 1, 1955; abstracted in *Econometrica*, vol. 24, no. 2, pp. 201–202, April, 1956.

Dorfman, Robert: "Application of Linear Programming to the Theory of the Firm, Including an Analysis of Monopolistic Firms by Nonlinear Programming," 98 pp., Bureau of Business and Economic Research, University of California, University of California Press, Berkeley, Calif., 1951. Reviewed by Erich Schneider, *Econometrica*, vol. 22, no. 1, pp. 129–130, January, 1954.

Frisch, Ragnar A. K.: "Principles of Linear Programming, with Particular Reference to the Double Gradient Form of the Logarithmic Potential Method," 219 pp., memorandum, University Institute of Economics, Oslo, Norway, Oct. 18, 1954.

Gunther, Paul: Use of Linear Programming in Capital Budgeting (letter to the editor), *Journal of the Operations Research Society of America*, vol. 3, no. 2, pp. 219–224, May, 1955.

Markowitz, Harry M.: "Portfolio Selection." Thesis submitted to the University of Chicago, Chicago, Ill., Summer, 1953; published in *Journal of Finance*, vol. 7, no. 1, pp. 77–91, March, 1952.

Martin, Alfred D., Jr.: Mathematical Programming of Portfolio Selections, *Management Science*, vol. 1, no. 2, pp. 152–156, January, 1955.

Samuelson, Paul A.: "Linear Programming and Economic Theory," *Proceedings of the Second Symposium in Linear Programming*, vol. I, pp. 251–272 (collection of papers presented to a conference sponsored by the National Bureau of Standards and the Directorate of Management Analysis Service, DCS/Comptroller, Headquarters USAF, held in Washington, D.C., Jan. 27–29, 1955); also P-685, The RAND Corporation, 17 pp., May 25, 1955.

———: "The Châtelier Principle in Linear Programming," 18 pp., RM-210, The RAND Corporation, Aug. 4, 1949.

Solow, Robert M.: Linear Programming: Lecture XII, "Notes from M.I.T. Summer Courses on Operations Research, June 16–July 3, 1953," pp. 116–129, Technology Press, M.I.T., Cambridge, Mass., 1953.

Whitin, Thomson M.: Classical Theory, Graham's Theory, and Linear Programming in International Trade, *The Quarterly Journal of Economics*, vol. 67, no. 4, pp. 520–544, November, 1953.

5. MILITARY APPLICATIONS

Jacobs, Walter W.: The Caterer Problem, *Naval Research Logistics Quarterly*, vol. 1, no. 2, pp. 154–165 (including references), June, 1954.

———: Military Applications of Linear Programming, *Proceedings of the Second Symposium in Linear Programming*, vol. 1, pp. 1–27 (including references) (collection of papers presented to a conference sponsored by the National Bureau of Standards and the Directorate of Management Analysis Service, DCS/Comptroller, Headquarters USAF, held in Washington, D.C., Jan. 27–29, 1955).

Joseph, Joseph A.: "The Application of Linear Programming to Weapon Selection and Target Analysis," 40 pp., *Operations Analysis Technical Memorandum* 42, Operations Analysis Division, Directorate on Operations, DCS/Operations, Headquarters USAF, Washington, D.C., Jan. 5, 1954.

Nicholson, George E., Jr., and George W. Blackwell: "Game Theory and Defense against Community Disaster," 71 pp., National Research Council, Washington, D.C., February, 1954. Reviewed in *Research Previews*, vol. 2, no. 3, pp. 1–5, May, 1954.

Saaty, T. L., and K. W. Webb: Sensitivity and Renewals in Scheduling Aircraft Overhaul, *Proceedings of the Second International Conference on Operational Research*, English Universities Press, Ltd., London, and John Wiley & Sons, Inc., New York, 1960.

Wood, M. K., and M. A. Geisler: Development of Dynamic Models for Program Planning, chap. 12 in T. C. Koopmans (ed.), "Activity Analysis of Production and Allocation," John Wiley & Sons, Inc., New York, 1951.

6. PERSONNEL ASSIGNMENT

Dantzig, George B.: "Notes on Linear Programming: Part XIV—A Computational Procedure for a Scheduling Problem of Edie," 13 pp., RM-1290, The RAND Corporation, July 1, 1954.

7. PRODUCTION SCHEDULING, INVENTORY CONTROL, AND PLANNING

Bellman, Richard E.: "Mathematical Aspects of Scheduling Theory," 61 pp., P-651, The RAND Corporation, May 23, 1955. Published in *Journal of the Society for Industrial and Applied Mathematics*, vol. 4, no. 3, pp. 168–205 (including references), September, 1956.

Cahn, Albert S., Jr.: The Warehouse Problem (Abstract 505), *Bulletin of the American Mathematical Society*, vol. 54, p. 1073, November, 1948.

Charnes, Abraham, and William W. Cooper: "Generalizations of the Warehousing Model," *ONR Research Memorandum* 34, Graduate School of Industrial Administration, Carnegie Institute of Technology, Pittsburg, Pa., also published in *Operational Research Quarterly*, vol. 6, no. 4, pp. 131–172 (including references), December, 1955.

———, William W. Cooper, and Donald Farr: Linear Programming and Profit Preference Scheduling for a Manufacturing Firm, *Journal of the Operations*

Research Society of America, vol. 1, no. 3, pp. 114–129 (including references), May, 1953.

———, ———, and B. Mellon: A Model for Optimizing Production by Reference to Cost Surrogates, *Econometrica*, vol. 23, no. 3, pp. 307–323 (including references), July, 1955. Also published in *Proceedings of the Second Symposium in Linear Programming*, vol. I, pp. 117–150 (including references) (collection of papers presented to a conference sponsored by the National Bureau of Standards and the Directorate of Management Analysis Service, DCS/Comptroller, Headquarters USAF, held in Washington, D.C., Jan. 27–29, 1955).

Efroymson, M. A., and T. L. Ray: A Branch-bound Algorithm for Plant Location, *Operations Research*, vol. 14, no. 3, May–June, 1966.

Fetter, R. B.: A Linear Programming Model for Long Range Capacity Planning, *Management Science*, vol. 7, no. 4, July, 1961.

Gepfert, Alan, and Charles H. Grace: Operations Research. . . . as It is Applied to Production Problems, *Tool Engineer*, vol. 36, no. 5, pp. 73–79, May, 1956.

Gomory, R. E., and B. P. Dzielinski: Optimal Programming of Lot Sizes, Inventory and Labor Allocation, *Management Science*, vol. 11, no. 9, July, 1965.

Johnson, Selmer M.: "Optimal Two- and Three-stage Production Schedules with Setup Times Included," 10 pp., P-402, The RAND Corporation, May 5, 1953. Presented to the Econometric Society meeting, Washington, D.C., Dec. 28, 1953; published in *Naval Research Logistics Quarterly* (Office of Naval Research), vol. 1, no. 1, pp. 61–68 (including references), March, 1954.

Kantorovich, L. V.: Mathematical Methods of Organizing and Planning Production, *Management Science*, vol. 6, no. 4, July, 1960.

Kelley, J. E., Jr.: Critical Path Planning and Scheduling: Mathematical Basis, *Operations Research*, vol. 9, no. 3, May–June, 1961.

Koeningsberg, E.: Some Industrial Applications of Linear Programming, *Operational Research Quarterly* (U.K.), vol. 12, no. 2, June, 1961.

Magee, John R.: Guides to Inventory Policy. Part I: Functions and Lot Sizes, *Harvard Business Review*, vol. 34, no. 1, pp. 49–60, January–February, 1956; Part II: Problems of Uncertainty, *ibid.*, no. 2, pp. 103–116, March–April, 1956; Part III: Anticipating Future Needs, *ibid.*, no. 3, pp. 57–70, May–June, 1956.

———: "Linear Programming in Production Scheduling." Paper presented to the first national meeting of the Operations Research Society of America, Washington, D.C., Nov. 17–18, 1952; abstracted in *Journal of the Operations Research Society of America*, vol. 1, no. 2, p. 76, February, 1953.

Manne, Alan S.: "An Application of Linear Programming to the Procurement of Transport Aircraft," 2 pp. P-672A, The RAND Corporation, May 13, 1955. Presented to the second national meeting of the Institute of Management Sciences, New York City, Oct. 20–21, 1955; abstracted in *Management Science*, vol. 2, no. 2, pp. 190–191, January, 1956.

Müller-Merbach, Heiner: The Optimum Allocation of Products to Machines by Linear Programming, *Fortschriftliche Betriebsfuhrung* (Germany), vol. 11, no. 1, February, 1962.

Rapoport, L. A., and W. P. Drews: Mathematical Approach to Long-range Planning, *Harvard Business Review*, vol. 40, no. 3, May–June, 1962.

Salveson, Melvin E.: The Assembly Line Balancing Problem, *Proceedings of the Second Symposium in Linear Programming*, vol. I, pp. 55–101 (including references) (collection of papers presented to a conference sponsored by the

National Bureau of Standards and the Directorate of Management Analysis Service, DCS/Comptroller, Headquarters USAF, held in Washington, D.C., Jan. 27–29, 1955); also published in *Transactions of the American Society of Mechanical Engineers*, vol. 77, no. 6, pp. 939–947, August, 1955; and in *Journal of Industrial Engineering*, vol. 6, no. 3, pp. 18–25 (including references), May–June, 1955.

Smith, S. B.: Planning Transistor Production by Linear Programming, *Operations Research*, vol. 13, no. 1, January–February, 1965.

Vazsonyi, Andrew: "A Problem in Machine Shop Loading." Paper presented to the fifth national meeting of the Operations Research Society of America, Washington, D.C., Nov. 19–20, 1954; abstracted in *Journal of the Operations Research Society of America*, vol. 3, no. 1, p. 115, February, 1955.

Whitin, Thomson M.: Inventory Control Research: A Survey, *Management Science*, vol. 1, no. 1, pp. 32–40 (including references), October, 1954.

8. STRUCTURAL DESIGN

Charnes, Abraham, and Herbert J. Greenberg: "Plastic Collapse and Linear Programming. Preliminary Report." Paper presented at the summer meeting of the American Mathematical Society, September, 1951; abstracted in *Bulletin of the American Mathematical Society*, vol. 57, no. 6, p. 480, November, 1951.

Dorn, W. S., and Herbert J. Greenberg: "Linear Programming and Plastic Limit Analysis of Structures," 30 pp. (including tables), *Technical Report 7*, Carnegie Institute of Technology, Pittsburgh, Pa., August, 1955.

Heyman, Jacques: Plastic Design of Beams and Plane Frames for Minimum Material Consumption, *Quarterly of Applied Mathematics*, vol. 8, no. 4, pp. 373–381, January, 1951.

9. TRAFFIC ANALYSIS

Lavallee, R. Stanley: "The Application of Linear Programming to the Problem of Scheduling Traffic Signals." Paper presented at the seventh national meeting of the Operations Research Society of America, Los Angeles, Calif., Aug. 15–17, 1955; abstracted in *Journal of the Operations Research Society of America*, vol. 3, no. 4, p. 562, November, 1955.

Little, J. P. C.: The Synchronization of Traffic Signals by Mixed-integer Linear Programming, *Operations Research*, vol. 14, no. 4, July-August, 1966.

10. TRANSPORTATION PROBLEMS AND NETWORK THEORY

Balinski, M. L., and R. E. Quandt: On an Integer Program for a Delivery Problem, *Operations Research*, vol. 12, no. 2, March–April, 1964.

Batchelor, James H.: A Commercial Use of Linear Programming, *Proceedings of the Second Symposium in Linear Programming*, vol. I, pp. 103–116 (including references) (collection of papers presented to a conference sponsored by the National Bureau of Standards and the Directorate of Management Analysis

Service, DCS/Comptroller, Headquarters USAF, held in Washington, D.C., Jan. 27–29, 1955).

Dantzig, George B., Lester R. Ford, Jr., and Delbert R. Fulkerson: "A Primal-Dual Algorithm," 16 pp., P-778, The RAND Corporation, Dec. 5, 1955. Also published as Part XXXI of "Notes on Linear Programming," 14 pp., RM-1709, The RAND Corporation, May 9, 1956.

———, and D. L. Johnson: Maximum Payloads per Unit Time Delivered through an Air Network, *Operations Research*, vol. 12, no. 2, March–April, 1964.

———, and J. H. Ramser: The Truck Dispatching Problem, *Management Science*, vol. 6, no. 1, October, 1959.

Dwyer, Paul S.: "The Solution of the Hitchcock Transportation Problem with a Method of Reduced Matrices," University of Michigan, Ann Arbor, Mich., December, 1955.

Flood, Merrill M.: Application of Transportation Theory to Scheduling a Military Tanker Fleet, *Journal of the Operations Research Society of America*, vol. 2, no. 2, pp. 150–162, May, 1954.

———: "On the Hitchcock Distribution Problem," 26 pp., P-213, The RAND Corporation, May, 1951. Presented to the Symposium on Linear Inequalities and Programming, Washington, D.C., June 14–16, 1951, jointly sponsored by the Air Force, DCS/Comptroller, Headquarters USAF, and the National Bureau of Standards. Published in Project SCOOP, Manual 10, pp. 74–99, Apr. 1, 1952; also published in *Pacific Journal of Mathematics*, vol. 3, no. 2, pp. 369–386, June, 1953.

Ford, Lester R., Jr., and Delbert R. Fulkerson: "Notes on Linear Programming: Part XXIX—A Simple Algorithm for Finding Maximal Network Flows and an Application to the Hitchcock Problem," 21 pp. (including references), RM-1604, The RAND Corporation, Dec. 29, 1955; also P-743, The RAND Corporation.

Fulkerson, Delbert R., and George B. Dantzig: Computation of Maximal Flows in Networks, *Naval Research Logistics Quarterly*, vol. 2, no. 4, pp. 277–283, December, 1955.

Gleyzel, Andre N.: An Algorithm for Solving the Transportation Problem (Research Paper 2583), *Journal of Research of the National Bureau of Standards*, vol. 54, no. 4, pp. 213–216, April, 1955.

Heller, Isidor: "Least Ballast Shipping Required to Meet a Specified Shipping Program." Presented to the Symposium on Linear Inequalities and Programming, Washington, D.C., June 14–16, 1951, jointly sponsored by the Air Force, DCS/Comptroller, Headquarters USAF, and the National Bureau of Standards; published in Project SCOOP, Manual 10, pp. 164–171, Apr. 1, 1952.

Hitchcock, Frank L.: The Distribution of a Product from Several Sources to Numerous Localities, *Journal of Mathematics and Physics* (Massachusetts Institute of Technology), vol. 20, no. 3, pp. 224–230, August 1941.

Kantorovich, L.: On the Translocation of Masses, *Comptes rendus (Doklady) de l'académie des sciences de l'URSS*, vol. 37, no. 7–8, pp. 199–201, 1942.

Koopmans, Tjalling C.: "Optimum Utilization of the Transportation System." Paper presented at the international meeting of the Econometric Society, Washington, D.C., Sept. 6–18, 1947; abstracted in *Econometrica*, vol. 16, no. 1, pp. 66–68, January, 1948; published in full in *Econometrica*, vol. 17, nos. 3 and 4, pp. 136–145 (with discussion, pp. 145–146), supplement to the July, 1949, issue.

Lederman, J., L. Gleiberman, and J. F. Egan: Vessel Allocation by Linear Programming, *Naval Research Logistics Quarterly*, vol. 13, no. 3, September, 1966.

Orden, Alex: The Transshipment Problem, *Management Science*, vol. 2, no. 3, pp. 276–285, April, 1956.

11. TRAVELING-SALESMAN PROBLEM

Dantzig, George B., Delbert R. Fulkerson, and Selmer Johnson: "Solution of a Large-scale Traveling-salesman Problem," 33 pp. (with references), P-510, The RAND Corporation, Apr. 12, 1954, revised July 8, 1954. Presented to the summer meeting of the Econometric Society, Montreal, Canada, Sept. 10–13, 1954; published in *Journal of the Operations Research Society of America*, vol. 2, no. 4, pp. 393–410, November, 1954; reviewed by H. W. Kuhn in *Mathematical Reviews*, vol. 17, no. 1, p. 58, January, 1956.

Flood, Merrill N.: "The Traveling-salesman Problem," 21 pp., Seminar Paper 13, Informal Seminar in Operations Research, 1954–1955, sponsored by The Operations Research Office and held at The Johns Hopkins University, Baltimore, Md., Feb. 16, 1955. Published in *Operations Research*, vol. 4, no. 1, pp. 61–75 (including references), February, 1956; also published in J. F. McCloskey, and J. M. Coppinger (eds.), "Operations Research for Management," vol. II, pp. 340–357 (including references), Johns Hopkins Press, Baltimore, 1956.

Heller, Isidor: On the Traveling Salesman's Problem, *Proceedings of the Second Symposium in Linear Programming*, vol. II, pp. 643–665 (including references) (collection of papers presented to a conference sponsored by the National Bureau of Standards and the Directorate of Management Analysis Service, DCS/Comptroller, Headquarters USAF, held in Washington, D.C., Jan. 27–29, 1955).

Little, J. D. C., K. G. Murty, D. W. Sweeney, and C. Karel: An Algorithm for the Traveling-salesman Problem, *Operations Research*, vol. 11, no. 6, November–December, 1963.

12. OTHER APPLICATIONS

Aronofsky, J. S., Growing Applications of Linear Programming, *Communications of the ACM*, vol. 7, no. 6, June, 1964.

Cohen, K. J., and F. S. Hammer: Optimal Coupon Schedules for Municipal Bonds, *Management Science*, vol. 12, no. 1, September, 1965.

Dantzig, George B., and Alan J. Hoffman: Dilworth's Theorem on Partially Ordered Sets, Paper 11 [pp. 207–214 (including references)] in "Linear Inequalities and Related Systems," *Annals of Mathematics Studies* 38, Princeton University Press, Princeton, N.J., 1956.

Ford, Lester, Jr., and Delbert R. Fulkerson: "Maximal Flow through a Network." Paper presented to the meeting of the Econometric Society, New York City, Dec. 27–30, 1955; published in *Canadian Journal of Mathematics*, vol. 8, no. 3, pp. 399–404, 1956; also RM-1400 and P-605, The RAND Corporation, 12 pp., Nov. 19, 1954.

Freeman, R. J., D. C. Gogerty, G. W. Graves, and R. B. S. Brooks: A Mathematical Model of Supply Support for Space Operations, *Operations Research*, vol. 14, no. 1, January–February, 1966.

Hartung, P. H.: Brand Switching and Mathematical Programming in Market Expansion, *Management Science*, vol. 11, no. 10, August, 1965.

Hoffman, Alan J., and Harold W. Kuhn: On Systems of Distinct Representatives, Paper 10 [pp. 199–206 (including references)] in "Linear Inequalities and Related Systems," *Annals of Mathematics Studies* 38, Princeton University Press, Princeton, N.J., 1956.

Knight, U. G. W.: The Logical Design of Electrical Networks Using Linear Programming Methods, *Institute of Electrical Engineers, Proceedings*, vol. 107, no. 33, June, 1960.

Kolesar, P. J.: Linear Programming and the Reliability of Multicomponent Systems, *Naval Research Logistics Quarterly*, vol. 14, no. 3, September, 1967.

Loucks, D. P., C. S. ReVelle, and W. R. Lynn: Linear Programming Models for Water Pollution Control, *Management Science*, vol. 4, December, 1967.

McGuire, C. B.: Some Team Models of a Sales Organization, *Management Science*, vol. 7, no. 2, January, 1961.

Mannos, Murray: "An Application of Linear Programming to Efficiency in Operations of a System of Dams." Presented at the summer meeting of the Econometric Society, Montreal, Canada, Sept. 10–13, 1954; abstracted in *Econometrica*, vol. 23, no. 3, pp. 335–336, July, 1955.

Orden, Alex: Application of Linear Programming to Optical Filter Design (abstract), *Proceedings of the Second Symposium in Linear Programming*, vol. I, p. 185 (collection of papers presented to a conference sponsored by the National Bureau of Standards and the Directorate of Management Analysis Service, DCS/Comptroller, Headquarters USAF, held in Washington, D.C., Jan. 27–29, 1955).

Wagner, H. M., R. J. Giglio, and R. G. Glaser: Preventive Maintenance Scheduling by Mathematical Programming, *Management Science*, vol. 10, no. 2, January, 1964.

Wardle, P. A.: Forest Management and Operational Research: A Linear Programming Study, *Management Science*, vol. 11, no. 10, August, 1965.

Wilson, R. C.: A Packaging Problem, *Management Science*, vol. 12, no. 4, December, 1965.

INDEX

{How index-learning turns no student pale,
Yet holds the eel of science by the tail}

ALEXANDER POPE

A CATALOG OF SELECTED
DOVER BOOKS
IN SCIENCE AND MATHEMATICS

A CATALOG OF SELECTED
DOVER BOOKS
IN SCIENCE AND MATHEMATICS

QUALITATIVE THEORY OF DIFFERENTIAL EQUATIONS, V.V. Nemytskii and V.V. Stepanov. Classic graduate-level text by two prominent Soviet mathematicians covers classical differential equations as well as topological dynamics and erqodic theory. Bibliographies. 523pp. 5⅜ × 8½. 65954-2 Pa. $10.95

MATRICES AND LINEAR ALGEBRA, Hans Schneider and George Phillip Barker. Basic textbook covers theory of matrices and its applications to systems of linear equations and related topics such as determinants, eigenvalues and differential equations. Numerous exercises. 432pp. 5⅜ × 8½. 66014-1 Pa. $8.95

QUANTUM THEORY, David Bohm. This advanced undergraduate-level text presents the quantum theory in terms of qualitative and imaginative concepts, followed by specific applications worked out in mathematical detail. Preface. Index. 655pp. 5⅜ × 8½. 65969-0 Pa. $10.95

ATOMIC PHYSICS (8th edition), Max Born. Nobel laureate's lucid treatment of kinetic theory of gases, elementary particles, nuclear atom, wave-corpuscles, atomic structure and spectral lines, much more. Over 40 appendices, bibliography. 495pp. 5⅜ × 8½. 65984-4 Pa. $11.95

ELECTRONIC STRUCTURE AND THE PROPERTIES OF SOLIDS: The Physics of the Chemical Bond, Walter A. Harrison. Innovative text offers basic understanding of the electronic structure of covalent and ionic solids, simple metals, transition metals and their compounds. Problems. 1980 edition. 582pp. 6⅛ × 9¼. 66021-4 Pa. $14.95

BOUNDARY VALUE PROBLEMS OF HEAT CONDUCTION, M. Necati Özisik. Systematic, comprehensive treatment of modern mathematical methods of solving problems in heat conduction and diffusion. Numerous examples and problems. Selected references. Appendices. 505pp. 5⅜ × 8½. 65990-9 Pa. $11.95

A SHORT HISTORY OF CHEMISTRY (3rd edition), J.R. Partington. Classic exposition explores origins of chemistry, alchemy, early medical chemistry, nature of atmosphere, theory of valency, laws and structure of atomic theory, much more. 428pp. 5⅜ × 8½. (Available in U.S. only) 65977-1 Pa. $10.95

A HISTORY OF ASTRONOMY, A. Pannekoek. Well-balanced, carefully reasoned study covers such topics as Ptolemaic theory, work of Copernicus, Kepler, Newton, Eddington's work on stars, much more. Illustrated. References. 521pp. 5⅜ × 8½. 65994-1 Pa. $11.95

PRINCIPLES OF METEOROLOGICAL ANALYSIS, Walter J. Saucier. Highly respected, abundantly illustrated classic reviews atmospheric variables, hydrostatics, static stability, various analyses (scalar, cross-section, isobaric, isentropic, more). For intermediate meteorology students. 454pp. 6⅛ × 9¼. 65979-8 Pa. $12.95